FORENSICS

CRIME SCENE INVESTIGATIONS FROM MURDER TO GLOBAL TERRORISM

DR ZAKARIA ERZINCLIOGLU

CONTENTS

LOCARD'S PRINCIPLE

By their fruits ye shall know them.
Matthew VII. 20.

The weather was cold and bleak, with a biting wind, which the sparse cover of the leafless autumn forest did little to check. I was standing by a tree, leaning on one of its low-slung branches and watching a team of policemen digging into the loose soil a few feet away. It was slow and tedious work for them, because, after every two or three spadefuls, I would ask them to stop, so that I could examine the soil they had placed carefully on the white sheet beside the ever-widening, ever-deepening hole.

Now, I asked them to stop again and I stooped to look at some of the soil with a magnifying glass. I took up the debris and placed it on an enamelled dish, examining the contents more closely. No, nothing there. Carry on, boys. They started digging again, but the days are short at the end of November and it was beginning to get dark. The men were getting tired. The digging had started early that day, but still nothing was found.

The reason we were not making good progress was that we simply did not know what we were looking for. The dead body of a murder victim had been found not far from this place and there were signs suggesting that it had been buried, but later exhumed and deposited in the place in which it was discovered. Soil and remains of leaf litter on the body indicated that the corpse may have been buried in the forest nearby and a search revealed a patch of ground where the soil had clearly been disturbed. Perhaps the body had been buried there. If this could be demonstrated, another piece of the chronological reconstruction of the crime would be accomplished. But the work had been in progress all day and still nothing was found.

At last, in the classic fashion of an unimaginative whodunit, just as the darkness threatened to call a halt to the work, something turned up. In fact, several things turned up all at once. The first was a human fingernail. The second, some strands of hair. The other things, less directly revealing perhaps, but informative nonetheless, were a few pupal cases of flies, which must have developed from maggots that had been feeding upon the corpse.

We now had evidence that a corpse had been buried in that spot in the forest, but was it the same body that now lay in the mortuary? We could not be sure, but it seemed very likely. Later, when the hairs from the burial site were compared with those on the dead body, they were found to match exactly. So that part of the mystery was solved.

Once the body was buried, it left its mark. The disturbed soil; the fingernail; the hair; and the tell-tale fly pupal cases were all the result of the burial. The clues left behind were the consequences of the act.

In 1921, some sixty years before these events took place, Sir Arthur Conan Doyle, creator of Sherlock Holmes and Doctor Watson, arrived at Lyons in France, on his way back to England from his travels in Australia. In Lyons he visited Dr Edmond Locard, one of the great names in forensic science, at his famous laboratory behind the Palais de Justice in the city.

Dr Locard was a great admirer of Conan Doyle. He always maintained that the Sherlock Holmes stories were of prime importance in the development of the science of criminal detection. Sherlock Holmes' use of apparently trivial clues, such as dust and cigar ash, were taken very seriously by him. In one of Conan Doyle's novels, Holmes remarked that he had published a monograph, *Upon The Distinction Of The Ashes Of Various Tobaccos*. In this little treatise, Holmes claimed that the examination of tobacco ash at the scene of a crime could furnish clues that would help to solve the mystery, a claim that many readers of the adventures of Sherlock Holmes still regard as very far-fetched. Yet Locard took a lead from Holmes and published a scientific paper on the significance of tobacco ash found at the scene of the crime.

He stated his opinion of the Holmes stories quite emphatically:

"I hold that a police expert, or an examining magistrate, would not find it a waste of time to read Doyle's novels … If, in the police laboratory at Lyons, we are interested in any unusual way in this problem of dust, it is because of having absorbed ideas found in Gross and Conan Doyle." *

Edmond Locard was delighted to meet Conan Doyle, whom he took on a tour of his Black Museum of criminal cases. He showed him the cabinets of weapons and all sorts of physical clues from the cases he had investigated throughout the world. He showed him the photographs of the various criminals who had been arrested by the application of Locard's own techniques.

Suddenly, Sir Arthur Conan Doyle stopped. He was looking at one of the photographs and he appeared puzzled. The photograph was of a young man with a moustache and cold, disdainful eyes. Turning to Dr Locard, Conan Doyle said: "But that is Jules, my former chauffeur!"

Dr Locard was astounded. "No, you must be mistaken, Sir Arthur, that is Jules Bonnot, the motor-bandit."

And so it was, but Conan Doyle was not mistaken. For Jules Bonnot – murderer, anarchist and terrorist – had, indeed, been his chauffeur in the years before the First World War.

Dr Locard told Conan Doyle the whole story of what was known of the life and death of Jules Bonnot. In one of those odd coincidences that prompt the exclamation that truth is stranger than fiction, it transpired that Bonnot had also served as the chauffeur of another celebrated criminologist, Harry Ashton-Wolfe, who worked as the assistant of Alphonse Bertillon, chief of criminal investigation in Paris. In 1911, some time after he left Sir Arthur's employ, Bonnot embarked on a new career of crime. Until that time, he and his associates had attacked the bourgeoisie by stealing their cars. Now, they began to use the car itself as a weapon, driving about the country committing robberies and murders. At last, with the forensic help of Dr Locard, Bonnot and his gang were brought to justice – violently.

In April 1912, the police besieged a garage in Choisy-le-Roi, near Paris. Inside it were Bonnot and the other members of the gang. Shooting began; one of the gang was killed and Bonnot was mortally wounded and, cursing his captors, was taken to hospital to die.

Dr Locard was horrified to hear that the creator of Sherlock Holmes had been driven around by the murderous Bonnot. Conan Doyle could have been murdered, he said. "Bonnot chauffeuring Sir Arthur Conan Doyle! The flesh creeps. Think how close we came to not having all we do have of Sherlock Holmes!"

Edmond Locard was credited with the success of many other police investigations, including the case of Emile Gourbin, which took place in the same year in which Bonnot was killed. Gourbin, a Lyons bank clerk, was accused of murdering his lady-friend, but he had what appeared to be an unshakeable alibi. When Locard examined the body of the young woman, he found marks around her neck, which indicated that she had been strangled. He took scrapings from under Gourbin's fingernails and found that flakes of skin that could have come from the woman's neck were present in the debris. But that was not all, for the skin flakes were coated with the same cosmetic powder that the victim had used on her neck and face. Confronted with this evidence, Gourbin made a full confession and was convicted.

Dr Locard came to be known as the Sherlock Holmes of France, a title of which he was very proud. He practised his skills for a very long time. Born in 1877, he studied medicine and law at Lyons, eventually becoming the assistant of Alexandre Lacassagne, a pioneer criminologist and professor of Forensic Medicine at the University of Lyons. He held this post until 1910, when he began the foundation of his criminal laboratory. He produced a monumental, seven-volume work, *Traité de Criminalistique*, and continued with his researches until his death in 1966, aged almost ninety.

EVERY CONTACT LEAVES A TRACE

What has all this to do with the excavation of a presumed burial site in an English forest so many years later? There is a link, albeit not a unique one. Edmond Locard, having absorbed the ideas of Lacassagne, Bertillon, Gross and Conan Doyle, was able to focus his thoughts and formulate the basic principle of forensic science – a principle so simple and, with hindsight, so obvious, that some might say that it is hardly worth spelling out or aggrandizing with the title of "Principle". Yet it is the basic idea upon which all forensic science is ultimately based, for Locard's Principle simply states: every contact leaves a trace.

The burglar who touches a window pane with his bare hands leaves his fingerprints behind; the prowler who steps on to the flower-bed leaves his tracks and carries some soil away on his boots; the robber who smashes a window carries minute fragments of glass on his clothes; the murderer may be contaminated by his victim's blood; the victim may retain in his hand some fibres from the murderer's clothes, pulled out during the struggle.

Yes, it is a simple idea, but, like many simple ideas in science, it is a very powerful one. It clears our thoughts and helps us to concentrate our efforts. Newton, observing the apple fall, came up with the simple idea, which seems so obvious to us now, that the Earth pulled it down. Darwin, studying the animals on the various Galapagos Islands, arrived at the simple conclusion that those individuals best suited to their environment would have a greater chance of survival. Archimedes, simply by sitting in his bath and watching the water rise, reached the simple conclusion that the volume of water that was displaced was equal to his own volume (or as much of him as was under the water). All very simple indeed.

Yet is Locard's Principle really equivalent to these other notions? Is it necessarily true that every contact leaves a trace? The critical reader might say that, although we found a fingernail and a few other things in the forest, it was quite possible that we could have failed to find anything. What if we had simply found nothing? What if the culprit had packed the soil more firmly, making it difficult to discern whether or not it had been disturbed? How could Locard's Principle have applied then?

The answer to this is that, every time a grasshopper jumps, the planet Earth is pushed in the opposite direction, but we may not be able to detect such a movement. By this I mean that Locard's Principle remains true, even if we cannot find a trace. The trace will be there, but the likelihood of our finding it will be limited by our abilities, our knowledge and the degree of refinement of the techniques and equipment at our disposal. Had we found nothing in the forest, chemical analysis of the soil may have revealed the presence of decomposition products from the body. The change in the composition of the microscopic flora and fauna in the soil may have provided a clue, and so on. The idea may be simple, the practical application of it may not be. The trick lies in actually finding the trace.

Formulating a principle, however simple or obvious, is a great boon, for it helps us to think clearly. When I taught students at Cambridge University, I used to tell them, before designing and carrying out an experiment, to write down the question they wanted to answer. However much difficulty they may have had at the outset, they were surprised at how much easier it was to design the experiment when they had actually formulated the problem in writing. They may have already "known" what the question was, but writing it down, as clearly as possible, crystallized it and made the problem much easier to handle. We seem to a great extent to think in language and in words.

Forensic science is concerned with finding out what happened in the, usually, recent past. It is akin to other areas of human endeavour, like history, archaeology and palaeontology, the aim of which is to discover the course of events that took place long ago. The time-scale may be much shorter in forensic science, but the thought processes involved are much the same. In fact, one could say that Locard's Principle forms the basic tenet of all these subjects.

I have always been fascinated by the past and how we can get to grips with it – how we can find out what really happened. This "past" may be the relatively recent past, as when one is conducting a forensic investigation of a murder; the more distant past, as in the reconstruction of historical events during the period of written history; the even more distant past, that which forms the subject matter of archaeology; or the remote past, the realm of the fossil-hunter. During the course of my professional career, I have had the opportunity – the privilege – to contribute to all these fields of study.

This book is about finding the traces that can lead us to discover the truth about past events – specifically, past events that are of interest to the law. Finding a buried skeleton that turns out to have belonged to someone who died thirty years ago – or even seventy-five years – is a matter for the law and the police, but the body of a person who died a hundred years earlier is of historical, not legal, interest. Forensic science and history merge at the edges, for where one ends, the other begins. Yet it is as well to bear in mind that forensic science has an applicability far beyond the area of the law, for its techniques and thought processes are widely used to interpret the facts – the traces available to us – in other historical subjects. These traces may not necessarily be physical ones – actual objects that one can hold in one's hand. They may be ideas that took shape, or events that took place, as consequences of earlier ideas or events.

You are standing by a pond and you throw a pebble in the water. There is a splash around the spot where the pebble entered the pond. Ripples are formed; they grow gradually fainter and less discernible towards the edges of the ever-widening circle. If someone else, standing beside you, throws a larger stone into the water, the ripples from it will collide and merge and mingle with the ripples from your pebble and it will be difficult to know the extent to which your action was responsible for the disturbance of the water. Once the pebble has disappeared beneath the surface, all we can observe are the tell-tale signs of the ripples. These ripples – these traces – will be our only clues to the past, so we must make the most of them, knowing that they could easily mislead us. Our task is not only to find these traces, but to find the correct interpretation of them and to attribute to them their true significance.

---

* Hans Gross (1847–1915) was the Austrian author of *System der Kriminalistik*, the first comprehensive work on forensic science, published in 1891. It was translated into English in 1907, under the title of *Criminal Investigation*.

CHAPTER ONE

THE FORENSIC MIND

*If a man will begin with certainties, he shall end in doubts; but if he will
be content to begin with doubts, he shall end in certainties.*

Francis Bacon

The Advancement Of Learning

I was preoccupied as I passed through the prison gates into the outside
world. The prison warder said good-bye and I answered him absent-
mindedly, for I was deep in thought. The conversation with the prisoner
was still fresh in my mind and I was troubled by it.

"I'm innocent, doctor! I didn't kill him! I've done many things – I'm a
crook and a thief, but I'm not a murderer!"

Melodramatic and predictable were the protestations of Jim, as I shall
call him; but they appeared quite genuine, nonetheless. I was inclined to
believe him and I suspected that he was unjustly imprisoned. Why? Why
did I believe him? The evidence against him was damning – he even
admitted that he had been in the house just before the murder was
committed. His hat was lying in a pool of blood. He had lied to the police
about his movements and he only told the truth later. He certainly was a
crook, a thief and a liar; and he was in prison for murder. The case against
him was very strong.

And yet...

I sat brooding in the car as the private investigator drove me home. It
does not fit, I told myself. What does not fit? I went over the facts once
again. Jim had gone into the house with another man, a well-known
ruffian. They threatened the occupant, a fellow crook who had double-
crossed them. The ruffian then pulled out a knife and held it up to the
man's face.

At this point fact and assertion become blurred. Jim says he panicked
and ran away, leaving the ruffian in the house; the police say that he helped
the ruffian kill the man. They invoke the evidence of the murdered man's
bruised wrist, which indicates that he was held by one man – Jim – while
the other man stabbed him in the neck. The jury were convinced and Jim
was convicted of murder and sent to prison for life.

Does this make sense? Does the evidence of the bruised wrist really indicate that two men were present at the murder? By the time I reached home, I had the answer. No, it does not make sense. In fact, the bruised wrist indicates that only one man was at the murder scene; it is evidence in favour of innocence, not of guilt.

Why? Consider this: if one man is holding a knife with one hand and trying to subdue another man with his other hand, he can only grip and bruise one wrist. If two men, with three spare hands between them, are trying to do the same thing, they could hardly fail to grip and bruise both wrists. What appeared, initially, to be evidence against Jim, became evidence in his favour.

SOUND REASONING

The practice of forensic science is not simply the application of a set of laboratory techniques; it is an attitude of mind, a tendency to think in a particular way. It is the acquisition of the habit of starting with a doubt, of being eager and willing to question the unquestioned. It is the cultivation of a suspicious mind.

Suspicion, of course, is of little use on its own. Hosea Ballou, the American founder of the Universalist Church, once said that suspicion is more often apt to be wrong than right, more unjust than just. Unfounded or emotionally based suspicion is, indeed, worse than useless and can cause great harm; it is what guides the lynch mob. For suspicion to be productive, it must be based on, and followed up with, sound reasoning. It has often been remarked that common sense is the one commodity of which everyone feels they have a sufficiency, yet common sense, paradoxically, can lead us astray in the most unexpected ways. What is "obvious" or what "stands to reason" is so often not reasonable at all.

This mixture of suspicion and reason is the forensic scientist's forte. It is not merely an interesting optional extra; it is essential. Without it, forensic science is reduced to the routine application of scientific recipes. The techniques of forensic science will be discussed in some detail in this book, but the thought processes that lead one to decide which technique is to be used, and in what way, are the subject of this chapter.

I have said that what seems reasonable or obvious may yet be totally wrong. The following example will demonstrate the truth of this, seemingly puzzling, assertion. Although it is not, strictly speaking, a "forensic" example, in that it is not concerned with a legal problem, it is a particularly good example of what I mean. Remember, too, that the techniques of forensic science are the techniques of reconstructing the past, whether that past is of legal interest or not.

El Greco, the renowned Spanish-Greek painter, adopted a most peculiar artistic style toward the end of his life. One does not have to be an enthusiast of renaissance art to be able to recognize a late El Greco painting on sight, for the human figures in them are disproportionately tall and thin. These features can be seen in almost all his late paintings and are particularly well depicted in such works as *St Martin And The Beggar*, in which the beggar is shown as a very attenuated figure. In *St Bernardino Of Sienna* the figure of the saint is similarly elongated. The angels in *The Adoration Of The Shepherds*, unlike the chubby cherubim of most Renaissance painting, are long and distorted.

These strange and unnatural depictions of many characters, including the figure of Christ himself, puzzled artists and scholars for many years, for El Greco was a consummate painter, who produced some well-proportioned portraits of important ecclesiastical figures of his day. Why, they wondered, did El Greco change his style in this strange manner? Could it be that he lost his sense of (literal) proportion as he grew older?

Some years ago, a distinguished eye surgeon suggested that El Greco painted his figures long and thin because, in his old age, he suffered from an eye defect that made him see people and things as long and thin. He painted them that way because he saw them that way. This seemed to make perfect sense and many people were satisfied that the mystery had been explained; old age and failing eyesight caused the bizarre change in style. It stood to reason.

But is this the true explanation? Think of what happens when one has a defect of vision. Let us say that an artist had a defect that made him see double. Every person or thing he looked at would appear to be two of the same thing. If he decides to draw a picture of, say, a man or a house, the man or the house will appear as two men or two houses. Every stroke of his pencil or brush will also appear double. Although he will draw a single man or a single house, it will appear double to him – and it will appear single to others.

If such a person should draw his figures double, he would see four figures in the end. This is not what he would normally see. It is *single* figures that he sees as double. Again, if he saw ordinary figures as being long and thin, he would draw ordinary figures, which would appear long and thin to him. If he should *draw* them long and thin, they would appear even longer and thinner, but only figures drawn according to their natural proportions would appear to him as the real objects would. So, he would not draw them abnormally long and thin.

No, El Greco did not have a defect of vision. Moreover, he *could not* have had one. The fact that his figures were long and thin is evidence for his *good* vision, not *vice versa*, as suggested by the eye specialist. El Greco painted his figures long and thin because that is how he wanted them to appear. The cause was voluntary, not involuntary.

This example shows how a causal explanation that may seem perfectly reasonable initially may actually be not only mistaken, but impossible. What causes what is often not easy to determine. Sometimes, certain things seem to go together, like heat and light. When we come across a source of cold light, we are mildly surprised. Yet, if we should come upon a hot object that emits no light, we are not nearly as surprised. Although we expect both to go together, we are happier with the idea of heat without light, rather than the other way round.

Of course, these expectations are the results of our experience. We know that the hot pan on which we burnt our hands does not emit light, but that most light sources we have touched are hot. Our general rule, then, becomes "no light without heat, but sometimes heat without light". It is a short step from this statement to the conclusion that heat is necessary to produce light – that heat is the cause and light the effect.

This is not true. Light is the visible part of the electromagnetic spectrum; heat is the energy that is transferred from one object to another as a result of a difference in temperature. There is, of course, a link between them; each is a kind of energy and we know that one kind of energy can change into another. Furthermore, one source of energy may emit that energy in more than one form.

The point here is that, although different things may go together – may be correlated, as scientists like to say – this does not necessarily mean that one causes the other. In a scientifically precise example like the one above this point can be seen fairly clearly; everything is accurately measured and we can readily see what caused, or did not cause, what. In dealing with evidence in situations that are not so easily measured the picture may become more confused. One piece of evidence that goes hand in hand with another may cause great confusion, not only in science, but in courts of law and in attempts to understand history, for correlations can be supremely and dangerously deceptive things.

Evidence that goes together with other evidence can confuse in many other ways. It is possible to present a piece of evidence in isolation, or in conjunction with another piece of evidence. In one kind of situation, presenting the evidence in isolation may be very misleading, yet in other situations presenting the evidence hand in glove, so to speak, with other evidence may be even more misleading.

Consider what your reaction would be if you saw a man walking along the street clutching a knife in his hand. Most people would feel that the man represented a threat; his appearance would seem menacing, even criminal. To put it more formally, the evidence of the knife would suggest that he was a dangerous man.

But what would your reaction have been if the man had been clutching a knife and a fork in one hand? Clearly, he would not have appeared in such an unpleasant light as he did when he was clutching only a knife. In fact, far from appearing menacing or criminal, he would have seemed amusing or even absurd. The evidence of the fork diluted the evidence of the knife, or more accurately, it modified the evidence of the knife. The fork made us look at the knife in a different way, with the result that our opinion of the man changed fundamentally.

Let us look at this man's behaviour more closely. We concluded that he appeared threatening when he was walking along with the knife on its own, but this does not necessarily mean that he was a criminal. It is our reaction to him, when he was carrying both knife and fork, that is really interesting. The fork may, indeed, suggest that the knife was an innocent object, but it may have been that very fact that made the man carry the fork in the first place; in other words, he may have carried it to deceive us into believing that he was harmless. On the other hand, if we had rushed to the police station and reported the sighting of a man walking about town with a knife, and withheld the information about the fork, the investigating police officer may subsequently feel that we had misled him by not mentioning that relevant little fact.

It is easy to think of various situations in which evidence, when taken together with other evidence, can give a very different picture from the one that would appear if the evidence had been looked at on its own. If the man in our example had been carrying a happy child in his arms, he would have looked quite harmless, even if he had been clutching the knife without a fork.

The opposite of the above is also true: evidence presented in isolation may lead us to very different conclusions from those we would have reached had we been given some other evidence at the same time. If a man, wounded and bruised, were to come to our doorstep and tell us, quite truthfully, that someone had assaulted him violently, we would feel that he had a valid grievance. However, we may look upon the situation very differently if we subsequently discover that our battered visitor had initiated the fight and that the beating he received was merely the result of the other man's defensive action.

CAUSE AND EFFECT

Perhaps these examples are too simple, too obvious, so let us take a more complicated example. When I used to teach at Cambridge University I often supervised groups of three or four students at a time. These supervisions involved discussing various scientific topics and I frequently discussed problems of logic and reasoning with them. In one such group, one of the students sometimes used to appear wearing a bright red jumper. The other students and I noticed that he always appeared quite cheerful when he wore this jumper and that he was never quite so pleased with himself when he was not wearing it. The link between his good spirits and his arrival in the red jumper was so strong that I felt moved to ask the group why they thought this was so. The chap himself was very good-natured and entered into the spirit of the thing, agreeing with my request that he should not divulge the true reason for this link, assuming there was one, until the end of the discussion.

His colleagues came up with a number of ideas. The first was that he wore that jumper whenever he felt happy, hence the correlation. No, our friend replied, this was not the reason. Someone else suggested that the cause and effect were the other way round – he felt happy whenever he wore the jumper; in other words, the jumper made him happy. Again, this was not the case. The jumper did not cause his happiness and his happiness did not make him wear the jumper; neither was the cause of the other. Neither the "chicken" nor the "egg" came first, yet the correlation between them was so strong. What, then, was the explanation?

The answer was simply that he wore the jumper when he went rowing. On those occasions when the supervision period immediately followed a bout of rowing, he would appear in my room wearing the red jumper. It was the rowing that caused the wearing of the jumper and the young fellow's happiness; both the jumper and his good spirits were the consequences of a cause that was invisible to us. The rowing "chicken" produced the two "eggs" of jumper and cheerfulness.

This kind of situation can arise very frequently in more complicated forms, often causing a great deal of confusion and misunderstanding. The reasonable conclusion that the red jumper and the good spirits were cause and effect, or vice versa, would have caused further confusion if the student had decided, halfway through term, to wear a blue jumper on his rowing excursions instead. We would then have been presented with the confusing fact of him being cheerful while wearing a totally different jumper. The point of all this is that we may often be presented with a number of effects and no causes, but that, in the absence of the latter, we tend to assume that one of the effects is the cause of the others. Such a conclusion would not necessarily stop there, since our reasoning might lead us on to actions or further arguments that would lead us deep into areas of utter confusion.

The following example will show how this kind of confusion can cause serious and even dangerous misunderstandings. It will also demonstrate that one false piece of reasoning can lead us into further errors. Someone once asserted in a magazine article that the presence of street lighting causes crime. He claimed that burglars and violent criminals were attracted to brightly lit areas, since such areas were usually inhabited by wealthy people. In a fit of hubris, he was so carried away as to assert that dark alleys at night were safer than brightly-lit thoroughfares. He supported these claims by referring to the true fact that there are more robberies and burglaries in brightly-lit suburbs than there are in suburbs that do not have street lighting.

There are two errors in this argument. The first error had to do with cause and effect, although the case was more straightforward than the red jumper story. The two observed facts – the bright lights and the robberies – were, indeed, cause and effect, but the writer got them the wrong way round, because this suited his purpose. (His purpose was made clear later in his article, but there is no point in going into that here.) In fact, bright lights do not cause crime. The known facts show that, when burglaries increase in certain residential areas, the residents demand better lighting in order to deter burglars. Burglaries drop as a result of the newly-installed lights, but still remain more common in such areas than in others. Therefore, the burglaries caused the lights, not the lights the burglaries. The second error was the "explanation" given. Having convinced himself that brightly lit areas are more prone to crime than others, the writer of the article then claimed that this was because criminals are attracted to bright lights. Since his first point was incorrect, his explanation of it was meaningless; one cannot explain something that did not happen.

Nevertheless, the fact that he did this shows how easy it is for one logical error to lead us into total confusion. Concerned that people may be misled by the article, with disastrous consequences, he felt obliged to publish a refutation of the article.

The unconscious selectivity of evidence can often lead us astray. Consider the following remark, variations of which are very commonly heard: "Yesterday I was thinking about Aunt Mary and, bless me, she telephoned that very afternoon. I do believe we are telepathic." The speaker and Aunt Mary may well share a telepathic sense, but Aunt Mary's telephone call is not evidence for it, for is it not true that Aunt Mary is thought about almost every day? The hundreds of occasions in which Aunt Mary came to mind, but after which she did not telephone, are forgotten; the single occasion in which she made contact is remembered.

In situations of this sort the evidence as a whole is not examined, only those parts of it that support a cherished belief. It is comforting that Aunt Mary should telephone when one thinks of her. This kind of mental trick, whether consciously or unconsciously adopted, is not confined to ordinary, everyday things, but affects our ideas in criminal trials and about historical events of all kinds. I have even heard eminent barristers argue their cases in court this way and, sadly and disturbingly, a number of recent political decisions were based upon such false reasoning.

TOO MANY ASSUMPTIONS

An example of recent political action of this kind concerns the ban on private ownership of handguns following the tragedy at the school in Dunblane, Scotland, in which a number of children were shot dead by a killer, Thomas Hamilton, with a handgun on March 13, 1996. Whether this ban was justified or not is a complex issue and cannot be dealt with summarily in a few sentences. My purpose is to examine the reasoning that led to the decision to impose the ban.

The ban was intended to prevent such an evil from happening again – or at least to minimize the chances of a recurrence. It was argued that if fewer people owned guns, the likelihood of anyone being shot by a gun would be reduced. If Hamilton had not owned a gun, he could not have committed the multiple murder. Furthermore, gun ownership is high in the United States, a country with a much higher murder rate than Britain, so it would seem reasonable to conclude that the prevalence of handguns in the population leads to more killings than would be the case if fewer guns were available. So far, so good.

However, this reasoning uses only part of the evidence available. For example, there is evidence to suggest that some of those parts of the United States that have the highest levels of gun ownership are blessed with the lowest crime rates in the country. Certainly, the crime rate in the United States is decreasing, while that in Britain is increasing, although it is still much higher in the States than it is in Britain. This fact may appear unpalatable, but it is a matter that should be considered. Furthermore, Switzerland has the highest gun ownership levels in Europe, yet it is the most law-abiding country, at least as far as violent crime is concerned. So, it is perfectly valid to ask whether decreasing gun ownership in Britain would have a beneficial effect. One may put the question more pointedly, albeit more crudely, and ask whether Hamilton would have dared to commit his crime had he not felt that no one was likely to fire back at him – the probability of anyone at the school owning a gun being very low. If would-be murderers knew that law-abiding citizens, too, possessed guns, would they be so ready to threaten others with their own? Is it not feasible to suggest that an increased ownership of guns by law-abiding people might act as a deterrent to violent criminals? I cannot answer these questions, since I do not have sufficient information upon which to base an opinion, my point is simply that the evidence suggests that these are valid questions that have to be addressed.

There are many other arguments, both for and against the ban on handguns. I do not wish to review these, since the point I wish to make is that choosing some evidence, simply because it supports one's emotional preference, and ignoring other evidence that is not sympathetic to one's preferred solution may lead people, and even governments, into taking the wrong decision, i.e. the decision that is less likely to bring about the desired change. Whether the handgun ban is justified or not, it was made on the basis of an emotional preference, namely, that the possession of a handgun is a bad thing. Deciding upon a course of action before examining all the evidence is a very dangerous procedure. In the case of the handgun ban, it is worth noting that murders are hardly ever committed by people who own such guns legally; Hamilton was an exception. Since most firearms murders are committed by people who own illegally acquired guns, the ban is hardly likely to make Britain a safer place. Also, Hamilton would not have been given a licence to own a handgun if the legislation existing at the time had been followed. If Hamilton had not been allowed a licence, either he would have not committed the murders, which is what the legislation was there to prevent, or he would have committed them with an illegally acquired weapon, in which case a ban would not have stopped him.

OCCAM'S RAZOR

At this point, I would like to introduce a half-forgotten figure from English history, who had some valuable things to say about reason and evidence. He is William of Occam, a Franciscan monk who lived during the late thirteenth and early fourteenth centuries. William taught that "entities are not to be multiplied beyond necessity". This turgid philosophical dictum can be expressed more usefully by saying that it is always best to consider the simplest explanation of any problem first. In other words, let us not complicate matters unnecessarily. This useful principle has become known as Occam's Razor, with which unnecessary "entities" can be shaved off. If one is faced with a problem it is best to try and explain it without recourse to too many assumptions. To take a simple example, let us say that you have returned home from work one day and found a book lying open on the table. One explanation is that another member of your family put it there. Another explanation is that it was a guest, admitted into the house by a member of your family, who opened the book. Yet another explanation is that a burglar gained entry and, when he had a free moment, decided to consult the book but left it on the table when he heard you coming. The first explanation is clearly the most likely. It may be said that this is, in fact, so obvious a conclusion that one does not need a mediaeval philosopher to draw one's attention to it.

But consider again. Although we know "instinctively" what the most likely explanation is, it is instructive to ask ourselves why we think that. The answer is that it is the explanation that makes the fewest assumptions about the situation. In other words, we have no reason to suppose, in the absence of other evidence, that either a guest or a burglar entered the house. To accept these other explanations would require us to assume things that we have no particular reason to believe. The first explanation does not require us to assume anything out of the ordinary; that is why we accept it in favour of the others. Of course, our settling on the simplest explanation does not mean we are necessarily right. If we find that we were mistaken, we may then consider more complicated answers, but it is always logical to start with the simplest answer to our question.

Again, it may be said that one can arrive at such conclusions without the aid of any grand philosophy. While this is true in many cases, the above hypothetical case is a very simple one. We could have refined the question and asked, at the very beginning, which member of your family put the book on the table. If the book was a dictionary, you would have been at a loss, since such a book does not suggest a specific user, unlike a book about, say, gardening or dogs. Such a situation would require very deep thinking about the most likely person to have used the book and it would have been not at all easy to decide, instinctively, who had consulted it. Moreover, the simplest explanation of the presence of the open book on the table would not have been so readily reached if your family had been away. Someone else would have opened the book and placed it on the table, but who was it? Again, the simplest answer in this situation is not so obvious. It will require a careful evaluation of the probabilities to work out which of the several possible scenarios is the most likely one, or the one that involves us in making the fewest assumptions. In real situations, when a legal or historical matter is being investigated, the problem may be quite complex; here William of Occam's insight can help us to think clearly and to dissect the problem more thoroughly.

FALSE ARGUMENTS

Sitting in court, listening to the proceedings of a criminal trial, I have often heard false arguments presented by barristers and accepted, in all good faith, by juries. Unfortunately, pointing out such errors of thinking does not always result in a withdrawal of the argument. One is hardly ever thanked for doing it. Some barristers are quite determined to win, regardless of the way in which the victory is achieved.

This brings me to a rather sensitive matter. In court it is often difficult to know whether a particular false argument is simply an honest mistake or whether it is an intentional manipulation of the truth. On the face of it, in a discussion of this sort, it may not appear to matter which of the two it is, as long as the illogicality of the argument is demonstrated. Sadly, it does matter. It matters because the empty rhetoric of a barrister of the kind who is impelled by a desire to win at all costs may sway a jury more certainly than logic. Consequently, the force of the logic used to refute the argument and expose the fallacy needs to be very clear indeed. A courtroom is often a field of battle. This chapter would be incomplete if we failed to look at the less savoury aspects of evidence presentation. It is, in any case, instructive to consider not only the ways in which things should be done, but the ways in which they should not be done. Little good would be achieved if it is claimed that the forensic mind is always perfect.

A particularly common error has to do with the evaluation of individual pieces of evidence against the accused. An excellent example of this sort of error took place during the notorious criminal trial of O.J. Simpson, the American football player, who was accused of murdering his former wife and her lover.

Evidence presented by the prosecution showed that blood found at the scene possessed blood-grouping characteristics that matched Simpson's blood. It was stated in court that only one in four hundred people had blood with such characteristics and that, therefore, this was strong evidence that the blood at the scene was Simpson's. The defence retorted by saying that a very large number of people, equivalent to the full complement of spectators at a Los Angeles football stadium, as they dramatically put it, would share the same blood characteristics. Therefore, they said, the findings could hardly be seen as being evidence against Simpson. In fact, they claimed that the evidence was, effectively, useless.

But was it? In principle, the answer to this question in the abstract must be "Yes" or "No", depending on the circumstances. If Simpson had been accused of murder solely on the basis of these findings, then they could not be seen as being evidence against him. For why select him, when many thousands of others could have been chosen with equal validity? On the basis of the findings, the blood could have belonged to any one of them, with equal probability in each case. Under such circumstances, the evidence would, indeed, have been worthless.

But the evidence did not come to light under such circumstances. There was already a *prima facie* case against Simpson – in other words, a certain amount of evidence against him already existed. To put it another way, the case against him could be presented in terms of odds. Let us say, for the sake of the argument, that, from what we already knew about Simpson (i.e. from the existing evidence and before assessing the blood evidence), the odds against him were three to two in favour of him being guilty. Without going into the mathematics of this matter, odds and probabilities can be calculated on the basis of existing evidence. So, we performed such calculations and came up with the odds of three to two; in other words, that the evidence suggests that he is more likely than not to have been guilty. If the calculated odds had been three to two against him being guilty (or three to two in favour of him being innocent, which is the same thing), then the evidence would suggest that he is more likely than not to have been innocent.

The point is that we start with certain odds. These are known as the prior odds. Now we come to the value of the blood evidence. The questions we have to ask are these: How does this new evidence affect the prior odds? Does it strengthen or weaken them? Does it strengthen or weaken the case against Simpson? Or does it not affect our beliefs at all?

In the case of the blood evidence, this new information certainly strengthens our belief in Simpson's guilt. Scientists would say that the posterior odds – the new odds, after the latest evidence had been taken into account – are greater in favour of guilt. So, contrary to the claims of the defence, the blood evidence was far from useless; it strongly affected our opinion about what happened.*

This point can be made more clearly by an imaginary story. You are walking along the street and you see, at a distance, a man entering a shop. He is tall and fair-haired and reminds you of one of your friends. Subsequently, someone else tells you that he saw such a man enter the same shop at the same time; he was closer to him and noticed that he wore gold-rimmed spectacles, much like the ones your friend wears. Yet another person says that the man wore a particular tie, spoke with a Scottish accent, had a voice of a particular quality and had a distinctive gait – all of which are attributes of your friend. As the evidence accumulates, you become more and more convinced that the man who entered the shop was, indeed, your friend. You come to this conclusion because of the multiplicity of the evidence, although each one of the man's attributes, taken in isolation, would not have constituted evidence of his identity, since millions of people are tall, millions of people are fair-haired and so forth. But not many people have the *combination* of attributes that your friend has, which is why you were finally convinced that the man and your friend were one and the same.

Although all this may sound perfectly obvious, the underlying point is, nevertheless, often not grasped (or is obfuscated) when evidence is given in court. Astonishingly enough, errors that may appear so simple and obvious in everyday life are often made in the courtroom without being noticed.

How can this happen? Here, I can only give my impressions, since it is almost impossible to collect hard evidence to support an explanation. We are dealing with the undivulged thoughts and reactions in people's minds and we cannot know what these are. In the case of juries, it is illegal to ask how members arrived at their conclusions. So, the answer to the question must, to a great extent, be speculative.

The atmosphere in a criminal court is always tense. Very serious matters are discussed. Someone's life, or at least future, is at stake. Barristers almost always exude an aura of great and ponderous authority and are usually endowed with enviable oratorial powers. They tend to be believed; few people think they would make such simple mistakes. Furthermore, the point, in a legal situation, is not as simply or as clearly manifested is it is in everyday situations, such as the case of the man entering the shop. Although the point in that hypothetical tale is exactly the same as the one made in the example taken from the O.J. Simpson case, most people would agree that, somehow, it was not quite as obvious as when the blood evidence was the point at issue. This may have something to do with the fact that scientific evidence tends to be presented and handled as though it were a special kind of evidence, to be assessed in a different way. People in general know that many men are tall, or fair, or wear gold-rimmed spectacles; but not many people know how many people have a particular blood-group. Even when the number of people having a particular blood-group is given in court, the new information is presented in a manner that makes it difficult for the lay person to assess its significance in the real world. Although I cannot speak for jury men and women specifically, scientific evidence seems to be regarded either with great reverence or great hostility by the population at large.

A SHIFT IN EMPHASIS

There are more subtle ways in which evidence can be misrepresented. Take the following example: say that a motorist is caught speeding by the police. He is stopped and breathalysed and found to be over the legal limit for driving.

Now, nothing devised by human beings is perfect and it may be that the motorist may register as being over the permitted limit, when, in fact, he was not. Let us say that the chance of this happening is very small; records may show that one per cent of the results of breathalysing tests are incorrect in this way.

In order to consider the driver's conduct in court, it will have to be stated that there is a one per cent chance of a positive result (i.e. that the driver had been over the limit) when, in fact, he had not been over the limit. This fact may be presented in court in a slightly different way: that there is a one per cent chance that the man was not over the limit, if he had got a positive result.

Are these two statements of probability the same thing? Although they look very similar, they are, in fact, not at all the same thing. Let us look at them again:

1. There is a one per cent chance of a positive result, when the driver was not over the limit.

2. There is a one per cent chance of the driver not being over the limit, if he got a positive result.

The statements are not at all the same, because the emphasis has been shifted from the probability of a positive result to the probability of the driver being over the limit. The difference may not be immediately obvious and it frequently goes unappreciated in court, especially when spoken by an able barrister. The shift in emphasis usually goes totally unnoticed, although that shift may make all the difference in the presentation of the evidence. The two statements are taken to be identical, whereas they are very different indeed, with potentially dire results.

An example from everyday life should make this point very clear. Consider these two statements:

1. All dogs are four-legged animals.

2. All four-legged animals are dogs.

I think most people would see instantly that these two statements are not the same, although exactly the same words are used in each one; only their order in the sentences is different. It is also clear that one cannot possibly deduce the second from the first. Although the error in our second example is blindingly clear, whereas it may not have been so in the first example, both are exactly the same kind of mistake, logically speaking.

Consider the following hypothetical example. If a defendant in the dock is found to have a certain type of handwriting and evidence is given that shows that, say, eighty-five per cent of criminals have handwriting with such characteristics, the court could easily be presented with a subtle shift in emphasis, thus:

1. Eighty-five per cent of criminals have type A handwriting,
could become,

2. Eighty-five per cent of those who have type A handwriting are criminals.

What was a statement of some interest has become a statement of strong probability of guilt. It is easy to see how an innocent man could be convicted on such evidence, simply because of this shift of emphasis. Other examples can easily be imagined. If you, the reader, think that this is too obvious a manipulation to pass unnoticed in court, unlike the drink and drive example, remember that the two alternative statements are not presented in court clearly and separately as they are above. Indeed, they are not presented as different statements at all, one simply changes into the other as the trial progresses.

A few years ago I was involved in the investigation of a murder in which a man was stabbed to death. His body was weighted down with bricks and deposited in a canal. I was asked to give an opinion as to the length of time the body had been immersed in the water. For various reasons, this proved very difficult to determine, although I could conclude that the body had been in the water for at least two days. It was possible that it could have been there for longer, but there was no evidence to suggest that it had been in the canal for more than the minimum two days.

When I attended to give evidence at the trial in the Old Bailey, the defence barrister asked me to confirm what I had said in my report, namely, that the evidence showed that the body was not in the canal for more than two days. I replied that I had not said this; what I had said was that there was no evidence to suggest that the body had been in the canal for more than two days. He retorted, somewhat testily, that that was the same thing, but, once again, I disagreed and pointed out that the two statements were very different. To my amazement, the barrister became somewhat sarcastic and made comments of a kind that suggested that I was either a pedant or a fool.

I had to explain that it was one thing to say that there was no evidence to suggest that the body had been immersed in the canal for more than two days and quite another to say that there was evidence that it had not been there for more than two days. To make the point even clearer, I said that the first statement meant that the body *may or may not* have been immersed for more than two days, whereas the second statement meant that the body had *not* been in the canal for two days. In my report I had made the first statement, not the second.

Astonishingly, he still could not (or would not) see my point and he became even more sarcastic, until the judge intervened. The judge had fully grasped the point and was able to make my meaning clear to the jury, but not before a great deal of confusion had been created in the courtroom. I can only guess what went on in the mind of the jury. As far as the barrister is concerned, I have often asked myself whether he really could not understand what I was saying, or whether he affected not to understand; but I have always puzzled over his apparent inability to grasp such a simple point.

Surprisingly, some barristers do seem genuinely incapable of understanding some very simple ideas, including concepts of justice. In my experience, some barristers, although sympathetic to the cause of some people, will refuse to act for them, because they disapprove of the group to which that person belongs. Examples of frowned-upon groups include politicians, extreme right-wingers, royalty and others. Such an attitude is far more damning than may appear at first sight, since we are entitled to ask of what other groups of the population do these barristers disapprove? In the present political climate it is easy to get away with condemnation of the groups listed above, but what if we should substitute for the above groups the following: blacks, homosexuals, old ladies or the unemployed? This would be considered an outrage and rightly so. People who are the victims of injustice are not necessarily people of whom one approves, but this should not affect one's view of the case. The fact that a barrister could not see this basic point is a cause for concern. I have often found myself in the position of giving evidence that would support the case of an individual whom nothing would induce me to befriend, but it never occurred to me that this fact should present me with a problem.

A VALID CONCLUSION?

As we have seen, it is often not possible to discern whether confusion or manipulation is the cause of the faulty presentation, so both have to be considered together. I should explain that I use the word "manipulation" to mean the presentation of evidence in such a way that weak evidence is made to appear stronger than it really is and strong evidence to appear weaker, with the purpose of influencing the jury's understanding of it. Much of what follows is critical of barristers, but I do not wish to give the impression that I am issuing a blanket condemnation of the behaviour of barristers in court. Many barristers I have known and worked with have been very open and straightforward in their handling of their cases and some shine as excellent examples of how to conduct oneself in a court of law. But I have also known barristers who were quite proficient at manipulating evidence in various ways.

One of the main sources of confusion or means of manipulation comes from the concept of "consistency". This, it has to be said, is a notion that lawyers seem to have acquired from forensic scientists, since it is often invoked by the latter when trying to explain the significance of their findings. Let me explain.

Consider the case of someone who was accused of breaking into a house and committing a burglary with violence. Glass fragments are discovered embedded in his shoes. Scientific tests on the glass show that it has properties similar to the glass of the window that was smashed and through which the burglar gained entry. As far as it goes, this is good evidence, as long as its strength is not exaggerated.

In court, a barrister, perhaps repeating the conclusion in the forensic scientist's report, may say that this evidence is consistent with the accused having broken the window. Although this is a true statement, it is nonetheless misleading, for it actually tells us very little, while having the effect of appearing to tell us a great deal. Very often, this kind of evidence is not discussed any further. The problem with it is that it does not tell us what else it is consistent with. In our example, the glass findings may have been consistent with the accused having broken any number of other windows with the same properties, but this is hardly ever pointed out.

There are worse examples. I have known several cases in which injuries to a child were said to have been consistent with sexual abuse, although they were also consistent with a dozen other possible causes. The point is that these other causes are not mentioned; only one is mentioned – the one that a barrister or a witness wants to implant in the jury's mind. There is no other reason for mentioning only one cause that is "consistent with" the evidence, otherwise why not choose at random any other cause? It is the choice of a particular cause and the with-holding of other possible causes that makes the pronouncement of consistency dangerously misleading. What should be done in such cases is to present all possible causes or scenarios that could have resulted in the evidence in question, be it an injury or the presence of glass in a shoe. The likelihood of each cause being the true one could then be argued and assessed in court.

Another kind of false reasoning can cause a great deal of trouble in court. Often a psychologist will report that a child who was interviewed exhibited certain attributes or mannerisms; say, he or she tended to avoid looking the psychologist in the eye or bit his or her nails while sitting and listening. The child may also have told the interviewer that he or she had dreams of a particular kind. Referring to the results from his own and other studies, the psychologist may find that eighty per cent of abused children exhibited the first mannerism, seventy-five per cent exhibited the second and ninety per cent had dreams of the kind described. He would then conclude that the child he interviewed was very likely to have been abused.

Is this a valid conclusion? On the face of it, it seems reasonable enough, but closer examination will show that it is deeply flawed. Without ever having examined an abused child, I can say, without fear of contradiction, that they all share another attribute in common: one hundred per cent of abused children breathe. This attribute – breathing – is even more strongly correlated with abused children than is any one of the other attributes that led the psychologist to conclude that the child had been abused. It is not how often a particular attribute is manifested in the behaviour of abused children that matters, it is whether such a characteristic is shared by other children as well. The question that should be asked is whether, and how often, children that have not been abused also show these characteristics. In the case of breathing the answer is perfectly clear, since we all know that all living children breathe. When it comes to mannerisms or dreams the answer is not so apparent; most people, including jury members, do not know how common such attributes are among children in general. The psychologist's conclusions, presented with statistical support, can appear to be rigorously scientific, when, in fact, they are not. The results and the conclusions present only half the picture; without the other half we cannot possibly arrive at a valid conclusion.

SCIENTIFIC METHOD

Contrary to some popular beliefs, science is a highly uncertain endeavour. It does not deal in certainties, but probabilities. Those accustomed to thinking of science as the production of high-tech equipment may be surprised by this assertion. In the minds of many non-scientists, "science" is computers, lasers and rockets to the Moon; but this is to confuse science with technology. While technological advances do, indeed, owe much to scientific research, the two activities are by no means the same thing.

Science is the activity concerned with the rational understanding of the natural world; technology is the production of machines or other processes designed to bring about a particular result. Technology uses science; and science uses technology. I mention this because many people look askance at the assertion that science is concerned with probabilities not certainties; perhaps they feel it is false modesty, or even an affectation, on the part of scientists. They will say that the television always works when it is switched on (unless it is broken); that the light comes on when the switch is flicked. Surely, then, science always works.

This is to misunderstand what scientists (as opposed to developers of technology) are trying to do. Scientists seek to explain why things are the way they are. Such explanations (or hypotheses) are put forward, tested as rigorously as possible and, if they withstand these tests, they are accepted as theories, until such time as they are shown not to work. In other words, scientists do not seek to "prove" theories – they do not believe they can do such a thing – rather, they fail to disprove them. Having bombarded an idea with arguments and experiments; and if the idea emerges unscathed, then it is accepted – for the time being.

So, science is concerned with probabilities at all levels of organisation. Physicists cannot say where a particular electron will be at a given moment – they can only say where it is most likely to be. Animal behaviourists cannot say which way the antelope will flee when it sees a lion – they can only suggest the most likely direction. Probability is a very common word in the vocabulary of scientists. Forensic scientists are no exception; they, too, must evaluate evidence on the basis of probability.

So, when interpreting the crime scene or the evidence taken from it for laboratory examination, one is looking, not for what happened, but what probably happened. A report is eventually produced that presents the results in terms of probability.

But, you may ask, the probability of what? The scenario, in other words, the mentally reconstructed idea of what *may* have happened is, strange to say, often quite different things in the mind of the police officer and in the mind of the forensic scientist. Police officers are apt to ask whether the story of the accused man is likely to be true, in view of the forensic evidence found at the scene. Forensic scientists ask the question the other way round; they ask whether the evidence is likely, in view of the story of the accused.

Who is right? Is it not all an exercise in semantics anyway? No, it is not; and the forensic scientist is right and the police officer is wrong. Before you say: "He would say that, wouldn't he?" or "The police officer is clearly the one with his feet on the ground," or "Trust a pedantic academic to come up with such nonsense!", consider the following.

The evidence at the scene is a fact; one either did or did not find a smoking gun on the floor. What may have happened is a hypothesis, an opinion, an idea – call it what you will – but is not a fact. So, if the investigating officer asks me whether it is likely that Bloggs was at the scene, in view of the fact that his gun was found there, my answer will have to be yes, it is likely.

How helpful is this? In truth, not much. One does not need a forensic scientist to tell one the obvious. But if the officer had asked me whether the evidence is likely, in view of Bloggs's story, I am able to say much more useful things. Why? Because I am no longer restricted to one possible scenario, namely, that Bloggs went to the scene and left his gun there. I would be left free to consider many other scenarios, many other possibilities and I would be able to assess, not only the possibility of each one, but to offer an opinion as to which one is the most likely – the one with the highest probability. Presenting the question the way the police officer did, leaves no room for manoeuvre – the answer to his question had to be yes, it is probable, but, much more dangerously, it closed the door that could have led to the consideration of other possibilities. It told us that it was, indeed, probable that Bloggs frequented the scene, but what else was probable would never have been known.

MATERIAL IRREGULARITY

In extreme, but not uncommon, situations the evidence could be equally consistent with guilt or innocence. Of course, the adversarial system allows the opposing barrister the opportunity to point this out, but the fact that a piece of evidence may be consistent with something else as well may not be apparent in many cases. This is the case when complicated scientific evidence is presented and for which no other explanation is immediately apparent to most people. Another explanation may well be apparent to a forensic scientist, who may choose not to reveal the fact. I regret to say that this does happen, as it did in the case of the Maguire Seven.

The Maguire Seven were arrested after they were suspected of the illegal possession of nitroglycerine. Swabs from the hands and scrapings from under the fingernails of all seven, as well as from gloves belonging to one of the seven, were taken and tested in the Royal Armament Research and Development Establishment (RARDE)** , using a technique known as thin layer chromatography. The results were said to indicate the presence of nitroglycerine in the samples. In other words, the results were consistent with the notion that the Seven had handled nitroglycerine.

The defending barrister questioned whether the results could have been caused by some other substance, but he could not suggest what such a substance could be. He also raised the possibility that the nitroglycerine contamination could have happened innocently, but he was unable to convince the court. The Maguire Seven were convicted in March 1976.

There is little point in rehearsing in detail what happened next. In essence, the case was investigated and it was found that other substances could have produced the same results as nitroglycerine and that innocent contamination, e.g. by handling nitroglycerine-contaminated towels, could have been an alternative explanation. It was also found that the scientists involved knew that other substances could have produced the observed results, but that they had withheld this information. The scientists had said that no further tests could be done after the initial tests, but the investigation showed that, not only could this have been done, but that it had, in fact, been done.

The Court of Appeal quashed the convictions and the six surviving members of the group (one had died in prison) were released. The Court considered that "material irregularity" had occurred due to the failure of the scientists to disclose all relevant information and concluded:

"We are of the opinion that a forensic scientist who is an adviser to the prosecuting authority is under a duty to disclose material of which he knows and which may have some bearing on the offence charged and the surrounding circumstances of the case. The disclosure will be to the authority which retains him and which must in turn... disclose the information to the defence... We can see no cause to distinguish between members of the prosecuting authority and those advising it in the capacity of a forensic scientist. Such a distinction could involve difficult and contested enquiries as to where knowledge stopped but, most importantly, would be entirely counter to the desirability of ameliorating the disparity of scientific resources as between the Crown and the subject."

However, the Court acquitted the scientists of any deliberate attempt to mislead the court at the original trial.

Interestingly, an almost exact parallel of this case took place at roughly the same period. In 1974, Judith Ward was convicted of bombing the National Defence College in Buckinghamshire, a coach on the M62 and Euston Station. Samples taken from Miss Ward's hands and from various belongings were tested by scientists at the Home Office Forensic Science Service, using a method known as the Griess Test, and by scientists at RARDE, using thin layer chromatography. Both laboratories concluded that traces of nitroglycerine were present.

After the acquittal of the Maguire Seven and the Birmingham Six, the Home Secretary referred the Ward case to the Court of Appeal, which quashed the convictions. As with the Maguire Seven, it found that "material irregularity" had occurred and that the scientists' reports were "calculated to discourage investigation". It transpired that substances from shoe polish could give the same results as nitroglycerine and that the scientists knew this, but withheld the information; that tests carried out at RARDE to consider the possibility of innocent contamination were misreported or withheld; that results recorded as "faint" were reported as "positive"; that some conclusions from the results were demonstrably wrong; and that medical evidence had been suppressed.

EXCESS OF ZEAL

In December 1972, a woman was raped and murdered in Scotland and her body was discovered in England some time later. For various reasons, a Mr John Preece was suspected of the crime; he was arrested and sent for trial in Scotland. Normally, the Home Office Forensic Science Service (HOFSS) would not have been involved in the case, its work being usually confined to investigations in England and Wales, but since the body was found in Cumberland, England, and the initial assumption was that the crime had occurred there, forensic scientists from the Chorley laboratory of the Home Office Forensic Science Service became involved. In particular, Dr Alan Clift, one of the laboratory's senior scientists, gave evidence at the trial, saying that blood belonging to the same group as Preece's blood was found on the dead woman's clothing. He also gave evidence concerning hairs that were recovered from the woman's collar and fibres from Preece's lorry.

Preece was convicted in 1973 by the High Court of the Justiciary in Edinburgh. The Chief Constable of Cumbria wrote a letter to the Director of the Chorley laboratory, in which he thanked Dr Clift for the work he had done and saying that his evidence had formed "the bulwark of the Prosecution's case". In February, 1974, John Preece appealed against his conviction, but the appeal was dismissed by the Scottish Court of Criminal Appeal.

In 1976, Dr Clift moved to the Birmingham laboratory of the HOFSS to take charge of its Biology Department and the following year new quality control procedures were implemented. During the course of his new duties, the Assistant Director came upon evidence that showed that Dr Clift had been selective in his reporting of the facts in court and had reported some facts incorrectly. In September, 1976, he concluded that Dr Clift was guilty of grave technical incompetence in the case of John Preece and in other cases and informed the Director of the laboratory of his opinion.

These discoveries were made when Dr Clift was away on leave. On his return, the Director and Assistant Director of the Birmingham laboratory interviewed him about a number of his cases in which his written reports were at variance with the recorded test results. As a result of this interview, it was concluded that Dr Clift had been selective in his reporting and that he had reported "results" that were clearly wrong.

Soon after this, a meeting of Home Office officials took place. It was attended by the Controller of the HOFSS and the Director of the Birmingham laboratory. The meeting concluded that the Director of Public Prosecutions would have to be informed in order to ascertain whether Dr Clift's actions were carried out with criminal intent. It was also decided to suspend Dr Clift from duty on full pay.

Following Dr Clift's suspension, the Director of the Aldermaston laboratory of the HOFSS, who had been asked to look at some of Dr Clift's earlier case work, sent a report to the Controller, commenting on the poor quality of Dr Clift's work and stating that he had made conclusions based on flimsy evidence; that he had failed to disclose the full facts; that he made "risky" decisions; that his conclusions were often "unsound and invalid"; and that his reports were sometimes ambiguous.

A police investigation into the conduct of Dr Clift took place, but the Director of Public Prosecutions decided not to prosecute him. However, in 1979, the papers relating to the forensic work in the Preece case were recovered from Dr Clift (they had gone missing from the Chorley laboratory) and early in the following year the Director of the Aldermaston laboratory of the HOFSS reported on the evidence in the Preece case. Her conclusion was that Dr Clift's interpretation of the fibre evidence was meaningless and, most damning of all, that the dead woman's blood group was the same as that of Preece; in other words, the blood on her clothes could equally well have come from the woman herself as from Preece. The evidence was neutral. The terrible thing is that Dr Clift knew this, but suppressed the evidence, knowing that his action would incriminate Preece.

The Controller, upon reading the Director of Aldermaston's report, nevertheless concluded that there had been no "excess of zeal" by Dr Clift. We shall return to this interesting phrase later.

In 1981, the High Court of the Justiciary in Scotland quashed Preece's conviction. It concluded that: "No reasonable jury would have convicted once it had become clear that Dr Clift was discredited not only as a scientist but as a witness upon the accuracy, fairness and objectivity of whose evidence reliance could be placed". The Parliamentary Ombudsman produced a report on the case, pointing out that bad forensic science evidence could lead not only to the conviction of innocent people, but also to false pleas of "guilty" by people who may come to believe that their case is hopeless. One might add that bad forensic science can also lead to the acquittal of guilty persons.

Why did Dr Clift behave in such a discreditable way? What did he have to gain? He had a secure job in a prestigious laboratory and had nothing to lose by speaking the truth, the whole truth and nothing but the truth. Many forensic scientists believe that cases of this sort arise out of a desire to please the police. There is much that supports this belief. Consider what happens when a police officer takes a case involving a serious crime to a forensic scientist. A certain amount of evidence attesting to the guilt of a particular individual is put before the scientist; all that is needed is to tie up some loose ends. The police officer may well say something like, "The man is dangerous, we have a strong case against him, but if we cannot clinch it he could easily go free. He is a rapist and a child-murderer. That poor girl would still be alive if it had not been for that man. Doctor, we would appreciate it if you could show that..." And so on.

This kind of subtle pressure and appeal to the emotions exerts a strong influence on some people. A forensic scientist may well come to feel that it is his duty to ensure that the monster who murdered an innocent child be brought to justice. The police, with their persuasive powers, have convinced him of their suspect's guilt: "There's no doubt about it, doctor, he is a brutal murderer and your evidence will be very important..." The scientist may well come to fear that, if he does not produce the necessary results, he will have been instrumental in the release of a dangerous criminal. Fear that he may be accused, however subtly, of allowing the man to go free may well influence his behaviour. It might lead to an "excess of zeal", to use the Controller's interesting phrase.

The use of this expression is, I believe, very revealing. Its very existence implies that it is a concept well known in the world of forensic science. Describing a forensic scientist as "zealous" suggests that he is being particularly helpful and conscientious, rather than dishonest and self-serving, which is what "zeal" really amounts to in a situation like this one.

I do not know what made Dr Clift behave in the way he did – I am simply saying that subtle pressure can, and does, influence the judgement of some forensic scientists.

But let us now look at the techniques at the disposal of the forensic scientist – the armoury itself. To pursue this military analogy, the choice of weapon will depend upon the nature of the problem. But it is also true that each problem brings with it special challenges to the forensic mind; and we will have occasion to learn more of the forensic mind as we pass through the arsenal.

---

*On October 3, 1995, Simpson was acquitted of murder at the end of a long criminal trial, but in 1997 a civil trial found him liable for the battery of his ex-wife and the "wrongful death" of her lover.

**In 1991 RARDE became part of the Forensic Explosives Department of the Defence Research Agency (DRA), which in turn became the Defence Evaluation and Research Agency in 1995, before being split in 2001 into the Defence Science and Technology Laboratory (DSTL) and QinetiQ.

CHAPTER TWO

SPIRIT OF PLACE

Not of the letter, but of the spirit; for the letter killeth,
but the spirit giveth life.
II Corinthians III. 6.

The chairs were overturned; the table was a mess; the floor was littered with all kinds of debris; rotting contents overflowed from the dustbin; and the smell was foul. A dead body was lying crumpled in a corner of the kitchen. I had just arrived and was surveying my surroundings as the CID officer gave me a brief sketch of events. I was standing at the scene – the scene of the crime, probably; at any rate, it was the scene at which the body was discovered.

The scene. It is a word that has a very special meaning to a forensic scientist. It evokes not a beautiful vista of sea or mountains, nor a panorama of lakes and forests. Rather, it conjures up a picture of tragedy, misery and squalor, for the "scene" is not scenic.

It is also, usually, a chaotic place. The police would have been the first to arrive there, alerted by a telephone call that something was amiss. What they do in those first few moments may affect everything that happens afterwards. The classic image of policemen trampling all over the evidence before the specialists are called is not wholly unfounded, but these days scene-of-crime officers, or SOCOs as they are called, take a look at the scene from the outset, with the aim of avoiding any such damage from taking place.

The SOCO takes a general look at the situation. The dead body clearly requires the attendance of a pathologist; the blood splattered on the walls and floor suggests that a specialist in these matters be called in; the maggots in the body point the need for an entomologist. Fingerprints, too will have to be sought and, if found, "lifted", to use police jargon. The important piece of equipment at this stage is the telephone. All the relevant specialists are briefed over the instrument and they arrive as soon as they can, the earlier the better.

Either the SOCO or a police photographer will have taken a number of preliminary photographs, starting from the outside of the house, working inwards toward the spot where the body is lying. But his task does not end there. He waits for the specialists, who will ask him to take further photographs of this or that. The SOCO prepares a diagram of the scene, makes notes, takes measurements, records the presence of items of interest, hopefully with the minimum of disturbance to the situation as he found it. Nowadays digital single lens reflex (DSLR) cameras are generally used in preference to conventional photographs. Digital photographs have the ability to be time and date-stamped, a useful feature for establishing a "chain of custody", which we shall return to later. Video-recording or, even more recently, 360-degree photography of the scene using a computer-controlled scanner, digital camera and laser rangefinder is becoming the norm. This latter device can take large numbers of photographs and measure millions of individual points and, by combining this laser data with photographs from multiple scan locations, a three-dimensional image of the scene can be reconstructed. However, it is important to note that this should be done in addition to, not instead of, "still" photography of individual items of evidence or specific areas of the "scene".

The specialists, or "experts", as the police, with touching faith, call them, start to arrive. What happens next? The textbook account will say that the expert will don protective clothing and, knowing what evidence to look for, will set to work at once, collecting items and placing them in polythene bags, labelling them, making supplementary notes and generally behaving as though the whole business were a mere mechanical activity, as though this scene was very much like the last, with the evidence obligingly obvious and just waiting to be collected, so to speak.

The reality is very different. A crime is not a laboratory experiment, with everything in its proper place and all the conditions known. Indeed, it is the task of the forensic scientist, faced with the end results of a series of events, to deduce what those conditions might have been. In this sense, it is the reverse process to a classic laboratory experiment. What, then, really happens at the scene?

EXPECT THE UNEXPECTED

At first, nothing. A good look at the scene, with the least disturbance to it, is essential. There then follows some talking, the scientist asking the police officer and the SOCO for further information about one or other aspect. The specialists will often consult with one another, coming to some agreement as to who should conduct an examination first. Should it be the pathologist? Or would that damage the possible fingerprint evidence that may be retrieved? Should it be the entomologist? Or would that make the pathologist's examination more difficult? Perhaps the blood-splattering specialist should go first, since the patterns on the floor would be trampled upon, if someone else were to go before him. And so on.

Things are often not at all what they seem and it is very easy to make incorrect interpretations as a result of preconceived ideas. To counter this tendency, there is only one thing that can and should be done; one must expect the unexpected.

I was once taken to task for making that remark. What an affectation! I was told. Only an academic could say something so silly! The unexpected is, by definition, unexpected, so how could one expect it? The answer is that one can quite easily expect things not to be what they seem to be, even if, at the outset, one does not know what the truth is. One can tell oneself that there may be more than one interpretation, even if one is, initially, totally at a loss.

Consider the following crossword puzzle clue: "More Work". The answer is a word of six letters. The devotees of crossword puzzles will have no difficulty in seeing that the answer may be not at all what it might appear to be. They will tell themselves that the answer may well have nothing to do with additional tasks to be performed, even though it may take them some time to discover what the clue really means. The answer, in fact, is: "Utopia".

The reason this clue presented a challenge is because the answer, while perfectly straightforward, was unexpected. If the answer was elusive it was because not all possible meanings of the two words were considered. The mind has a tendency to select the evidence most in keeping with its expectations and to ignore the evidence that is not.

It may be protested that all this is very "clever", but that a crossword puzzle clue is nothing like a real-life situation. This is not true, since misinterpretations and misunderstandings are very common in every day life. In this context I am reminded of the story (whether apocryphal or not) of what happened when escalators were first introduced into the London Underground. It is said that, in the early days, few people used them and the authorities wondered why this was so. Eventually, someone realized what was wrong: it was a sign that read "Dogs must be carried". The problem was that few people had the necessary dog!

If it turns out that this amusing story is, in reality, apocryphal, the following one is not. Moreover, it is a story in which I did the very thing that I have been warning against. This took place when I was invited to give a lecture to an amateur natural history society in London. The subject of the lecture was "Forensic Entomology", the subject concerned with the use of insects as evidence in criminal investigations.

After I arrived at the venue for the meeting and was arranging my notes and slides, an elderly gentleman came up to me and asked what forensic entomology was about. I was a little surprised by this question, since I thought that the answer was obvious. I did not like to say so, however, and the gentleman then went on to ask whether it was to do with the use of insect biology in the investigation of crime, or whether it was the study of beneficial or legally-protected insects, such as bees or butterflies, that had been illegally killed. It had never before occurred to me that any other interpretation of the words "forensic entomology" was possible, but I could see that it was my familiarity with the subject that led me to make that mistake.

A MISCARRIAGE OF JUSTICE

To return to what may happen at the scene, consider the following case. A woman was found murdered in her house on a Thursday morning; she had been beaten to death. It was known that she had returned home from work at about 6 pmon Tuesday evening. Her husband did not have an alibi for the period between 6 pm and 8 pm and suspicion fell on him. The prosecution alleged that the murder took place some time during those two hours on Tuesday, since, when her body was discovered, the woman had still been wearing her formal clothes and her work badge. (She worked in a department store.) Evidence had been presented in court that she was in the habit of removing her badge and changing into casual clothes as soon as she returned home. This suggested that she was murdered a little after 6 pm.

I was not involved in this case from the outset; I only became involved after the husband was put on trial, convicted and sent to prison. I was asked by some friends of the convicted man to look into the forensic evidence, since they were convinced of his innocence. I read all the available documentation, both scientific and legal, and examined the photographs taken at the scene. It immediately became clear that a central piece of evidence from the scene had been overlooked, presumably because its significance was not grasped. If it had been, the whole reconstruction of events would have had to be revised.

The documentation revealed that several people testified that the woman was wearing a white blouse on Tuesday afternoon. The photographs of her body showed her wearing a black blouse, with the work badge appended to it. Since it is hardly likely that she changed her clothes, removing the white blouse and putting on, uncharacteristically, a formal black blouse, on Tuesday evening, then appended the badge once again after she changed, her death must have happened on the following morning, Wednesday. However, it was known that she did not go to work on Wednesday, so the conclusion one has to arrive at is that she was killed just before she was about to leave for work that Wednesday morning, not on Tuesday evening. The badge was appended because she was about to go to work, not because she had just returned from work. I can see no other possible explanation.

It is worth mentioning, as a matter of interest, that the husband did have an alibi for Wednesday morning. In spite of the emergence of this new evidence, he has not been released from prison.

This example shows how things may not be what they seem to be at a scene. No-one considered the evidence of the blouse; everyone was concerned wholly with the evidence of the badge. The result was a miscarriage of justice.

It is for reasons like this that flat-footed rules of scene investigation can be harmful. Of course, one has to work in a methodical and systematic way as far as the physical handling of the evidence is concerned, but there can be no mental or intellectual equivalent to that process.

EXAMINING THE SCENE

Having said that, it is worth going over the general methodology of scene examination. It hardly needs saying that every care should be taken not to disturb or contaminate evidence; and that written and photographic records are made throughout the investigation. Generally speaking, and quite logically, the scene is recorded from the outside inwards. In other words, if the body is found inside a house, notes and photographs of the building are taken first; then a closer study of the exterior is made, followed by an examination of the entrance hallway and, eventually, to the body itself.

To say that evidence must not be mishandled or contaminated may seem so obvious as to hardly need saying. However, with the best will in the world, it is very easy to contaminate evidence. In one case I know the pathologist was asked by a police officer whether he had noticed a certain object in one corner of the "scene" room. The pathologist said that he had not seen it and, moreover, he had not been to that part of the room during his visit. It was only when a police officer was able to show that the pathologist's fingerprints were all over the wall in that precise corner, that the pathologist had to admit that he had, in fact, been there! This was an honest mistake, but the story demonstrates the need for extensive note-taking; dependence on memory is very bad practice.

Another true story concerns a case of death by shooting, in which the body of a man was found lying at one end of a room. All the evidence suggested that it was a case of suicide, but the gun was found on a mantelpiece at the other side of the room. The wounds of the deceased were such that death had to have been instantaneous, so the man could not have shot himself then walked to the opposite end of the room to die. It was decided to treat the matter as a case of murder. This was called off, however, when one of the officers admitted to having moved the gun in case it got in anyone's way!

A most important matter associated with scene investigation is the continuity of evidence. Any mistakes made during the initial examination of the scene will affect the rest of the investigation, so it must be absolutely clear where, when and how a particular item of evidence was taken; and by whom. The evidence must be packaged and labelled in such a way that there can be no doubt as to its "history". This "chain of custody" is maintained by keeping a record of signatures on the label on the container – whether it be a bag, box, glass tube etc – when it is handed from person to person. Thus, the SOCO signs the container when the evidence is taken and packaged; when he hands it to a forensic scientist, both people sign and date it. When the scientist returns the item to the police, it is again signed by both parties; and so on. This ensures that the court will be satisfied that the piece of evidence being discussed at the trial is, in reality, the same as the piece of evidence taken from the scene. This does not, of course, guarantee that the evidence has not been switched at some point, but it makes it a good deal less likely. At any rate, it is the most that can be done.

In a case of murder or suspicious death the first person to examine the body is usually a police surgeon or other doctor, not a forensic pathologist. It is the task of the doctor to establish whether the person in question is, indeed, dead. In many, perhaps most, cases this will be perfectly obvious, but in some cases it will not. It is not unusual for people to be taken for dead and their bodies treated accordingly. My own maternal grandfather was rescued from the Mediterranean Sea when the ship on which he was sailing sank. Those who rescued him could detect no sign of breathing and they presumed he was dead. It was only later, when he began to revive and move, that the mistake was noticed.

The practice of forensic science often gives one an insight into the way people behave. I know of a case in which a man was presumed dead and was taken to the mortuary. A visiting forensic scientist noticed that the man was, in fact, alive and he informed the mortuary attendant. The attendant thanked the scientist and said he would take the necessary steps. The next morning the forensic scientist received a telephone call from the attendant, saying that he had now dealt with the matter; he had placed the man in the freezer overnight and that he was now truly dead. This is not an apocryphal story.

To return to the scene: the police surgeon will examine the body, perhaps feeling for pulse. Then a stethoscope will be used to detect any sounds of breathing and of heart beating. Contrary to popular images, the doctor will not use a mirror or a feather to determine whether there is life in the body; there is no alternative to a proper medical examination.

A doctor will arrive at any scene of death in order to certify that the person is actually dead. In most cases the matter will be quite straightforward. The police will be involved only if the doctor notices anything that arouses suspicion. Of course, suspicion does not always arise at this point; in most murder cases the police arrive before the doctor, because they are alerted by someone, or something, else. But, in cases that are not at first regarded as suspicious, much depends on the certifying doctor's ability to recognize suspicious signs and to notify the police.

AROUSING SUSPICION

What is a suspicious sign? As we have seen, there are no rules for this kind of thing. Some signs are so obviously suspicious that there is no difficulty at all in recognizing them. An empty bottle of poison lying beside the bed is an example. However, there are more subtle signs; the doctor might feel that the body is not lying in an altogether natural manner. Inquiries may reveal that the deceased had collapsed and died and fell to the floor, whereupon his family picked up the body and placed it on the bed, showing that there was an innocent explanation. In other cases there may be no convincing explanation for the body's unnatural position. The doctor may notice a bruise on the deceased's neck, or what may appear initially to be a minor injury on the head. Were these caused by foul play, or can they, too, be innocently explained? Much depends on the doctor's ability and readiness to look for such signs; in the words of one forensic pathologist, the late Professor Keith Simpson, "the doctor is the watchdog of the public, and must keep an ever-open eye for the kinds of death that require an explanation".

In cases in which suspicion is aroused, the doctor must refer the matter to the Coroner without delay. The referral is usually made through the police. The Coroner's duty is to order an investigation. However, there are some cases of suspicious death – or "sussy" deaths, to use the somewhat grotesque patois of forensic pathologists – that are automatically regarded as suspicious; these are the sudden deaths of young people, previously not known to have suffered from a life-threatening illness. Such a death is treated with suspicion from the outset and is investigated accordingly.

I have often felt it odd that so many people seem to be unaware that post-mortems are obligatory in cases of suspicious death. This point was made very well in the television production of an Inspector Morse story, "Service Of All The Dead", by Colin Dexter. The victim was found dead, with a knife in his back, but forensic examination revealed that he had received an overdose of morphine as well. The murderer had drugged his victim, unwittingly using a lethal dose of morphine, with the purpose of changing his clothes (this is much easier to do with a living person than a dead one), then stabbed him. The forensic findings were important in the story, but the point here is that the culprit did not realize that a post-mortem examination would take place.

Removing a body from the scene is not always a straightforward matter. During daylight hours a group of ghoulishly interested by-standers may collect outside the door of a house and the policeman outside has the duty of persuading them to maintain a respectful distance. The body itself may not be easy to move without damaging some of the evidence, not least that concerned with the very position of the body when it was found. Inevitably, lifting the body and moving it to the ambulance will alter the position to a greater or lesser extent. The chalk line drawn around the body before removal, often featured in crime films, is what actually happens, wherever this is feasible. (It is not always feasible, however, for instance when a body is lying on a crumpled-up white sheet.)

The body is usually removed to a mortuary or to a hospital equipped to deal with post-mortem examinations. The work of the pathologist and other specialists at this stage will be discussed in later chapters; for the present we will concern ourselves only with the scene.

And yet, again, what exactly is the scene? Is the scene simply where the body was found, or is it where the murder actually took place? The two are not necessarily the same. Very often, a body is removed by the murderer from the place in which the deed was committed. An important part of the investigation must be to find out whether the body had been moved from another place and, if so, where that other place might be.

There are two questions here; we will start with the simpler one – has the body been moved? After death, when the blood stops circulating and, under the force of gravity, sinks to those parts of the body that are lying lowermost, causing purplish patches to appear. This phenomenon is known as hypostasis. So, if one finds a body with purple blotches on that side of the body that is uppermost, one can conclude that the body had been moved after death.

LOOKING FOR CLUES

Like so much else in criminal investigation, one has to look for clues wherever one can find them; the following is a case in point. The murdered body of a woman was found in a moorland and, for various reasons, the police wanted to know whether the murder had taken place at that spot, or whether it had been committed elsewhere, the body being dumped on the moor later. The woman had an injury to the head, which would have bled profusely. (Wounds to the head usually result in much bleeding.) Therefore, if the woman had died on the moor, there would have been a great deal of blood in the soil beneath her head. The forensic scientist consulted on the matter took two squares of turf from beneath the head and another two squares from the ground some distance away from the body. Chemical testing revealed the extent of blood-staining in the first two squares. Animal blood was added to the uncontaminated pieces of turf until the extent of staining matched that in the samples taken from beneath the body. This enabled the investigators to determine how much blood had soaked into the ground. The answer was about half a pint – a good deal, in other words. It was concluded that the murder must have taken place on the spot, since a body dumped there from another place could not possibly have bled to that extent.

Certain changes that take place when a body is placed in a particular environment can be used to determine not only whether a body has been moved, but from where it has been moved. For example, a body that had been lying in water is easily recognizable as such, even if it is found on dry land. Many of the changes we notice in ourselves after having a bath also occur in a dead body. These include the "goose-flesh" appearance of the skin, which is caused by the cold making the muscles that erect the small hairs on the body contract. In water, the head sinks low, causing the blood to gravitate toward the head and neck, making it more likely for decomposition to start in those areas.

Adipocere formation is another tell-tale sign that the body had been immersed in water or in a very moist place. Adipocere is a waxy substance formed by the action of water on the fatty acid, known as oleic acid, in the tissues. The "hydrogenation" of this acid produces stearic acid; it is essentially the conversion of an oily substance into a harder, more "fatty" substance. It is the same process that turns vegetable oils in margarine during the manufacturing process.

A mummified body, i.e. one that has dried out with the minimum of decomposition, indicates that it had been lying in a dry place. Mummification takes place because the dryness prevents, or at least greatly retards, bacterial decomposition. Dry conditions, aided by air currents, such as one might expect in a chimney or in the desert, make mummification much more likely. The bodies of new-born babies tend to mummify more rapidly than those of adults, because they are, to all intents and purposes, free of bacteria.

It is clear that these post-mortem changes can indicate that a body had been moved from a certain type of locality, but they cannot, in themselves, tell us what the actual geographical locality was. Depending on the circumstances of the case, certain actual localities may be suggested, such as when a body showing adipocere formation is found in a house not far from a lake. Clearly, the lake would seem to be the most likely place in which the body had been immersed, but the changes themselves do not allow that conclusion to be reached.

Happily, there are methods that can narrow down the number of likely places in which a body had been lying. There are no easy solutions in forensic science and it is worth repeating that one must look for whatever clues come to hand. Take the following case, which makes this point rather well.

A girl was found murdered in a garden. She had been raped; and a Potentilla leaf was found inside her underwear. The curious thing was that there was no Potentilla plant near the body, but there was one in another part of the garden. This suggested that the girl had been raped there, but that she was murdered as she was leaving the place. In fact, it transpired that the murderer had allowed the girl to dress after he raped her, then killed her as she started to leave. Although no-one could possibly have foreseen the presence of the leaf, it was just the sort of clue one should be on the look-out for. In many cases the insects that infest a body after death can give very valuable clues to the place of death. Insects have known habitat preferences and known geographical distributions, so it is often possible to arrive at conclusions about the kind of place (woodland, grassland, indoors, etc.) in which the body had been previously lying, as well as the part of the country in which it had been.

An interesting example of this occurred when I was consulted about the murder of a young man, whose body had been discovered in an upstairs flat. On his body were the larvae of a fly known as a dolichopodid (there is no common name), which breed in water or wet soil. It is inconceivable that such larvae could have bred in a human body at all, let alone one lying so far from water. There could be only one conclusion – the body had lain at the edge of some water and was removed to the flat later. The water-body most likely to have been the site of the murder was a large pond down the road. The lack of adipocere formation or any post-mortem change associated with immersion in water suggested that the body was lying close to the edge of the pond, but not in it.

I was consulted on the question of time of death in the famous case of Karen Price, the girl whose skeleton was found wrapped in a carpet and buried in the garden of a house in Cardiff. The insect fauna on the body was mostly of a kind that one would expect on a buried body, but there were also some bluebottle pupal cases. These revealed that the Karen's body had lain exposed above ground for some hours before burial, since bluebottles have no means of reaching a buried body. Moreover, despite the fact that they can easily detect a buried body, they will not lay eggs on the surface of the soil in the "expectation", so to speak, that the hatching maggots will burrow through the soil to reach the body. Some flies can do this, but not bluebottles.

Buried bodies have a very characteristic kind of associated insect fauna, which varies according to soil types. A body that had been buried, then exhumed and placed above ground, will carry with it the tell-tale signs of burial. I have had many cases of this sort, in which the presence of a typically subterranean fauna on an exposed corpse revealed the truth. One very common indicator of previous burial is a minute fly, known, quite appropriately, as the Coffin Fly. It hardly ever fails to turn up on buried corpses, whether or not they were coffined.

CORPUS DELICTI

Even an incompetent investigator is bound to find some clues at the scene. The real challenge comes when there is, initially at least, no scene and no body. John George Haigh, known to history as "Acid Bath" Haigh, claimed on his arrest that he could not be found guilty, since the police would not succeed in finding the body.

"Mrs Durand-Deacon no longer exists," he boasted. "She has disappeared completely and no trace of her can ever be found again."

When Detective Inspector Webb asked him what had become of her, Haigh replied:

"I have destroyed her with acid. You will find the sludge which remains in Leopold Road. Every trace has gone. How can you prove murder, if there is no body?"

Like others before him, Haigh had heard of the legal dictum of *corpus delicti*, which means "the body of the crime". The "body" referred to here is the body of evidence, not the body of a human being. Long before Haigh's time, people have been found guilty of murder without the discovery of a body; James Camb was found guilty of the murder of Gay Gibson, whose body he tipped out of a porthole into the sea.

Even among murderers, Haigh must be reckoned a very strange man indeed. He used to drink a glass of his victims' blood after murdering them; another one of his favourite beverages was his own urine. He enjoyed other drinks, too. After drinking a victim's blood, he would have a cup of tea – accompanied by a poached egg on toast.

Haigh robbed his wealthy victims and put their bodies in a tank of sulphuric acid on the premises of his plastic fingernail-making business. He had already murdered five people, before he committed the murder that would end his career. On Friday February 18, 1949, he invited Mrs Olive Durand-Deacon to visit his factory. She was never seen again; and her friend, a Mrs Lane, was worried. She told Haigh that she was going to the police, whereupon he replied: "I'll come along to the police station. I might be able to help." In the words of Professor Keith Simpson, the pathologist who worked on the case, it was "an outsize understatement".

The officer put on the case was a woman police sergeant named Lambourne, who instantly distrusted Haigh. Inquiries into his background revealed that he had a record of theft, fraud and shady dealing, but he had never been suspected of violent crime. After his boast to Inspector Webb, Professor Simpson was asked to pay a visit to the "factory", which was, in fact, nothing but a warehouse. On arrival, he looked at the rough, pebbled ground outside the warehouse and, almost instantly, he stooped and picked up what looked like one of the pebbles.

But it was not a pebble; it was a gallstone; laboratory examination subsequently confirmed that it was a human gallstone – an important point, since animals get them, too. Mrs Durand-Deacon had not disappeared without trace after all. Professor Simpson, ever ready to give credit where it was due, pronounced his immediate discovery of the gallstone as "impressive"; and I suppose it was.

Further examination of the site revealed some human bones, as well as a set of dentures. These last were shown to Miss Helen Mayo, Mrs Durand-Deacon's dentist. She had no doubt; they were the very dentures she had made for Mrs Durand-Deacon. The case against Haigh was complete. He was put on trial, convicted and hanged at Wandsworth Prison.

The case of "Acid Bath" Haigh has often been used to make two points, neither one of which is strictly, or wholly, true. First, as we have seen, it is said that someone can be tried for murder in the absence of a body. While true in principle, in fact this very rarely happens, the Camb case being one of the notable exceptions. I have known a case in which the body was stolen subsequent to my post-mortem examination; the case never came to trial.

The second point is that it is very difficult, if not impossible, to get rid of a dead body. Again, this is true in the sense that it is certainly no easy matter, but statements of this sort are, in fact, back to front – we know only about those bodies that have been discovered, despite all attempts to destroy them, but we know nothing about those bodies that were never discovered. After all, the bodies of Haigh's first five victims were never found.

Acid is not the only agent that has been used by murderers to dispose of their victims. Quicklime is another such substance, much favoured by writers of whodunits, although, in truth, quicklime is quite useless for the purpose. Lime is an oxide of calcium, having the chemical formula, CaO. When it comes into contact with water it turns into calcium hydroxide, with the formula, $Ca(OH)_2$. This compound, known as slaked lime, is indeed very corrosive, but its effect on buried bodies is not at all what one would expect.

When a body is buried and covered in quicklime, the oxide quickly reacts with water in the soil and in the body itself, turning into the corrosive slaked lime. Some corrosion of the body surface will take place, but the heat from the reaction is so intense that the body becomes quickly dehydrated – mummified, in effect. Mummified bodies, as we have seen, will not easily decompose.

The notorious nineteenth-century murderer, Henry Wainwright, who shot his mistress, Harriet Lane, in 1874, tried to dispose of her body by burying it in half a hundredweight of another form of lime, known as chlorinated lime. The body was so well preserved as a result of this treatment that it was quite easily identified when it was exhumed. Wainwright was hanged.

As a personal aside, it is said that Wainwright liked to eat walnuts and that he especially favoured those from a tree at Mile End in London. There is still a very old walnut tree there; and, being partial to walnuts myself, I have eaten many nuts from it. Someone once suggested to me that Wainwright might have been induced to commit murder by some constituent of the nuts. I am sure he was joking.

NEW TECHNIQUES

In cases like that of "Acid Bath" Haigh, there is no body, but there is a scene. In other words, there is a fairly well-defined area in which one could look for clues. But what can one possibly do if there is no body and no scene? What if someone disappears and is presumed murdered? Where does one begin?

Clearly, such a situation presents considerable practical difficulties, although, in the eye of the law, this does not necessarily mean that no conviction of a suspect can be made. Very frequently, the mysterious disappearance of a person can be linked to the activities of some other person. The police will make inquiries about the missing person by interviewing friends, relations, work-mates and others, one of which, for one reason or another, may well come to be suspected by the police. Nevertheless, it is must be remembered that many people disappear without trace every year and are never found. It is said that about twenty-five thousand such people go missing annually in the United Kingdom alone.

Once the police have a suspect, they will begin by searching his house and garden; as well as any other places he might frequent. Is the body anywhere in the vicinity? This is the time for heavy spadework, quite literally. Although there is no substitute for digging, many new forensic techniques have been developed that narrow down considerably the area that needs to be searched.

An example of such a method is ground-penetrating radar. A short pulse of electromagnetic energy is transmitted into the ground and, if there is an object present in the soil, a characteristic reflection will be received. This happens because the buried object will cause an alteration in the electrical properties of the soil. The information appears on a monitor that shows the pattern of the pulses. This technique was used successfully in the notorious Frederick West case in Gloucester, England, where the buried bodies of nine murder victims were recovered.

In this case the technique saved a great deal of time, effort and money, but the location of the bodies was never really in doubt. When a very large area is concerned, such as an area covering, say, half a county, other techniques have to be used. Aerial reconnaissance is very useful in such cases. It is much easier to discern a pattern on the landscape from a distance, much as one would "see" a painting better if one stood some feet away from it. A feature visible from the air may not be visible on the ground. Aerial reconnaissance is particularly valuable if older aerial photographs of the area are available, since these could be compared with the newer photographs taken during the investigation; and any differences in the landscape, such as vegetational or structural differences, may help to focus the search on an area that seems to have been disturbed. Spy-plane and satellite photographs were used to detect the presence of mass graves in Bosnia-Herzegovina, in particular the mass graves at Srebrenica.

Aerial photographs for forensic purposes are much more useful if they can be taken from different angles, in order to exploit the fact that differences in shading may highlight certain features. Also, regardless of shadows, some surface features may be more easily visible from one angle than from another.

Various methods of enhancing the contrast between different areas of the landscape have been in use in geological surveying and for military purposes for many years, although their application to forensic investigations is very recent. Thermal imaging is one of these. This technique makes use of the fact that the heat lost from disturbed soil (such as a burial pit) and that lost from undisturbed soil is different. A buried, decomposing body would release more heat than the surrounding soil. Clearly, this would help to narrow down the area of, say, a large field that needs to be investigated.

Multispectral imaging is based on the fact that different materials absorb and reflect the different bands in the electromagnetic spectrum differently. This information can be presented visually on a monitor and it is possible, at least in principle, to use this technique to locate spots on the landscape whose spectral pattern is not explicable on the basis of the known characteristics of the area. Although this method has long been used in the study of the fabric of old buildings, especially churches, its application to the analysis of forensic photographic and video evidence is much more recent with the advent of full-spectrum cameras and specialized image-processing software. The principle of a full-spectrum camera is that it utilizes light not just from the visible spectrum but also from the infrared and ultraviolet wavelengths that are invisible to the human eye. Soil disturbance in particular can be detected using these infrared and utraviolet spectra. These techniques are also being investigated with regard to enhancing images of archaeological sites.

The methods of geophysical prospection, long used by geologists and archaeologists, have also been applied to forensic science. One method involves the measurement of soil resistance. Here, the investigator pushes two metal probes into the ground, then passes an electrical current between them. The resistance is measured; it will be lower in moist soils than in drier, stonier soils, since water is a good conductor of electricity. The characteristics of a grave can be detected, since the resistance of the soil in it will be different from the surrounding soil. This resistance will often be lower than the surroundings, since the grave is likely to be more moisture-retaining than the adjacent areas, but it may sometimes be higher if the grave is filled with rubble or stones and if the body itself is wrapped, say, in a polythene bag.

Magnetometry is a technique that detects local changes in the Earth's magnetic field. A magnetometer is a device carried at a constant height above ground; it can register magnetic differences in iron-bearing soils. If the soil at a particular spot contains a buried body, it will contain fewer magnetic particles than the surrounding soil, thus registering a difference on the magnetometer.

These methods, which detect differences in the electrical resistance or magnetic properties of different areas of the ground, are not metal detectors as such. The classic metal detector of the amateur treasure-hunter is an electromagnetic device that can detect the presence of metallic objects, such as coins, buttons and watches. The finding of such artefacts, if modern, may give an indication as to whether a body is in the vicinity, although the presence of such objects does not, of course, mean that a body is lying buried in the area. In any case, the object being sought may not necessarily be a body; it may be a stolen item of jewellery, for example.

It is an interesting – and useful – fact that buried objects do not soon disappear into the structure of the subterranean soil and stones, as is generally supposed. The surface of the Earth is formed of stratified layers, although these layers may often be disturbed by human activity, which itself creates new layers. A buried body will create a new such layer, which will persist – and remain detectable – for a very long time. Another interesting fact that archaeologists often point out is that an object discovered buried very deeply in the soil is not necessarily older than one that is found nearer the surface. Areas that have been intensively used be human beings may have relatively recent layers lying very deep in the ground; conversely, some very ancient layers in areas that have seen very little human activity may be lie very near the surface.

Also, although it is generally true that murder victims are usually buried in very shallow graves, this is not always the case by any means. Long after the disappearance of the princes in the Tower of London in 1483, Sir Thomas More wrote an account of the reign of King Richard III, the purpose of which was to show how wicked the usurper king had been. Although More's account is full of errors and inaccuracies, it does contain one very startling assertion – startling, because of a discovery that was made a century and a half later.

Sir Thomas More wrote that the bodies of the two boys were buried "*at the stair foot, meetly deep in the ground under a great heap of stones*". In 1674, workers, digging down into the foundations of the White Tower at the very staircase indicated by More, discovered the skeletons of two children. They were found ten feet below the surface; "*meetly deep*" Sir Thomas More had written. If these were indeed the bodies of the two boys, then it is clear that a deep burial does not necessarily mean it is an ancient one.

Searching for a buried body requires more than a set of sophisticated techniques. At any rate, the search investigators should be very flexible in their approach, using whatever methods or evidence come to hand. In Colorado, USA, an interesting series of experiments was carried out some years ago. Using buried pig carcasses, various methods were brought to bear to discover their whereabouts. The study showed the need to use a multi-disciplinary approach in the search for concealed bodies. As so often in forensic science, there are few easy routes to the answer; it is as Thomas Edison said: genius is one per cent inspiration and ninety-nine per cent perspiration.

In the case of Stephanie Slater, who was abducted by Michael Samms in 1992 and kept alive for eight days in a cold coffin in Newark in Nottinghamshire, the ransom for her safe return was found buried in a field in Lincolnshire. The hoard of bank notes, totalling about £150,000, was eventually discovered, after a painstaking investigation. The known movements of the suspect, eyewitness accounts, the advice of archaeologists and the military, and even psychological reports, were all used to narrow down the area of search. When the "suspect" field was identified, the cache was found using ground-penetrating radar. Miss Slater was later released unharmed.

High-tech methodology is not always used in the search for bodies; in fact, it is still the norm not to use them. Such techniques are expensive to use and are conducted by highly paid practitioners, so the initial stages, at least, of a search are usually conducted using low-budget techniques. In any case, high-tech methods are not always applicable to the actual circumstances of a case.

Dogs are, famously, used to find both living people and dead bodies, buried or exposed. They can even detect bodies that are lying underwater and are routinely used in some parts of the USA for this purpose. During the Arab–Israeli Wars, dogs were used to find the bodies of soldiers buried up to one metre (more than three feet) deep. The olfactory abilities of dogs, although well known, are actually much more developed and impressive than is usually supposed, as we shall see in chapter three.

BURIED BODIES

Let us consider what one might actually do when searching for a buried body in a field. What tell-tale signs would one expect to find? Most soil is covered by vegetation of one kind or another and this would have been disturbed by the digging of the grave, so it makes sense to look for such disturbed areas. However, the burial may have taken place some time earlier and the vegetation may have covered it once again. But would it be the same vegetation? The presence of a body beneath the surface will encourage the growth of some plant species and retard the growth of others, since some plants prefer a highly fertile soil while others do not. Also, the high moisture content above the grave would also favour some species at the expense of others. Alternatively, if the buried body had been wrapped in polythene wrapping, or covered with stones, the moisture content may well be depressed, a fact that would be reflected in the type of vegetation above it.

More often, it is not so much the species of plant, but the vigour of vegetational growth, that will reveal the site of a grave. Again, one would expect a stronger vegetation growth above a body buried unwrapped and covered only with soil; while a wrapped body placed beneath stones will result in a less vigorous growth. Either way, any change of growth vigour or species composition from the general surroundings will indicate that the spot had been disturbed and was worth investigating. Differences in plant growth may be seen in drought-afflicted areas, where the area above a shallowly buried and uncovered body will reveal a much lusher plant growth. Such changes, however, are not always easily discernible from the ground; as we have seen, aerial reconnaissance is more likely to reveal such differences, but, in practice, climbing a tree and examining the surroundings from that vantage point is often very effective. Examination of such sites when the sun is low (either rising or setting) is more likely to show up the differences more clearly, because of the shadows that are cast.

When a grave is dug, much more soil is removed than can be put back in, since the body itself will fill much of the hole. Therefore, the surrounding area is likely to be covered, to a greater or lesser extent, by the removed soil. Also, the deeper the grave, the wider the adjacent area of disturbance is likely to be, since there will be much more soil lying about. Footprints and other marks may be discernible around the grave. In order to address this problem of excess soil, the perpetrator might compact the refilled soil with foot or spade, but this in itself will bring about a difference in vegetation growth.

Usually, a small mound of earth is apparent over the grave. With time, this will be compacted and sink, to become a depression. Compaction will be greater in deeper graves, resulting in deeper depressions. Another event that causes compaction is the decay of the body itself and the collapse of its skeletal structure. Indeed, two depressions, separated in time, may occur; one due to the compaction of the soil and another due to the collapse of the abdominal cavity of the body. The impression left is of a smaller depression within a larger one. Moreover, when the compaction of the soil takes place, cracks will appear at the edges of the grave where it meets the undisturbed ground – another sign of the presence of a pit.

The examination of such sites with artificial light in the dark might make it easier to detect them, because of the shadows that will be cast. Snow-covered ground is particularly revealing of irregularities.

When searching for a burial place all sorts of factors and kinds of evidence have to be taken into account. For example, geological records of the area to be searched may reveal that large tracts cannot hold a buried body, since the bedrock is shallow and very hard, confining the probable site of burial to a more limited section of the suspected area. The kind of soil trapped in the tyres of the suspect's car should reveal valuable clues. Gullies and ravines on the landscape are usually worth searching, since they are often used to bury bodies, because their depth allows easy burial. This fact was used to good effect during the investigation of the Moors Murders.

It is by no means easy to construct and camouflage a hastily and clandestinely dug grave. The perpetrators of crimes almost always try to dispose of the bodies of their victims in the easiest way and in a place that they believe is unlikely to be examined. For example, the bodies of many murder victims are concealed beneath the floorboards of houses, presumably in the belief that no-one would think of looking there. In fact, it is the first place one looks – I have looked under more floorboards than I care to remember. The attraction of the floorboards is that they are usually invisible in most modern houses with fitted carpets; this, together with the presence of tables, chairs and other furniture, tends to make the space under the floorboards appear remote and, as a hiding place of a body, unthinkable.

Hiding a body indoors has the advantage of allowing the perpetrator to work undisturbed, so bodies buried under the floorboards are usually, but not always, very well buried. I have known cases in which the body had been placed directly under the floorboards – in such cases the odour, if nothing else, betrays the crime. In other cases, the body is not simply deposited in the space beneath the floorboards, but under the concrete base that lies about a foot or so below them. The concrete has to be broken first, of course; then a pit is dug, the body buried and, finally, new concrete put in place. Even these measures fail to deceive, this method of concealing a body being so common.

EXCAVATING THE GRAVE

One extreme case of this sort of thing came my way when I was consulted about the disappearance of two lovers, Hashmat Ali and Sharifan Bibi. The couple had been cohabiting in a manner that incurred the severe disapproval of their families. At first, it was thought that the couple had gone to Pakistan, but inquiries revealed that this was not so; and it soon became obvious that they had been murdered.

Suspicion fell on the girl's two brothers. A house that used to belong to the family was examined and it was found that the flagstones in the cellar had been concreted over. Removal of the concrete revealed that the flagstones had been disturbed and broken. A bad smell emanated from the floor; and excavation work revealed the presence of a pit. This was systematically investigated and, although no bodies were found, some suggestive items came to light. Much as in the case at the very beginning of this book, a finger nail was recovered. More significantly, soap and ice-cream wrappers were found. These had on them the batch numbers that could be dated by the manufacturers, thus narrowing down the time of death.

The above work was carried out by specialist forensic archaeologists skilled in the proper excavation of graves. My own involvement came when I was asked to examine large quantities of the rather heavy clay soil removed from the pit, with the purpose of finding any living things that might shed light on the crime. Alas, after several days of very messy work involving the use of strong jets of water to fragment the clumps of soil. nothing was found. This may seem a somewhat sterile (metaphorically and literally) end to the story, but it serves to show the reality of forensic investigation – sometimes a great deal of hard work comes to nothing. This does not mean that the investigation, as a whole, failed. Far from it; the two brothers were tried for murder and convicted.

The excavation of a grave has to be carried out carefully. Until the very recent past, suspected graves were dug, rather than excavated, by policemen with spades and shovels, resulting in the destruction or loss of much of the evidence. Nowadays, archaeological methods of excavation are employed. It is important to limit the excavation to the genuine space of the burial pit, recognized by the difference in soil structure and compaction. As we have seen, the edge of the dug-out area of soil is often visible when the soil settles with the passage of time. This "cut" may not be easily discernible, since hastily-dug graves almost always have a jagged outline, making careful excavation more difficult. If digging is carried out indiscriminately, ignoring the delimitation of the dug area, extraneous evidence may be dug up, confusing the investigation. For example, a button may be unearthed; if it had come out of the genuine grave space, it would be evidence relevant to the case, but if it had come from that part of the soil outside the confines of the pit, then it is clearly not relevant, or at least highly unlikely to be relevant. At any rate, it is essential to document exactly where the item came from, so that a proper reconstruction can be made.

People's physical abilities and psychological tendencies must also be considered when searching for a clandestine grave. For example, most people cannot carry a dead body very far. A corpse is usually heavy and unwieldy, as well as being something that people generally do not want to carry. This means that corpses are disposed of as soon and as easily as possible. It is a good deal more likely to find a buried body downhill than uphill, simply because people find it much easier to dump a body down a slope, rather than carry it up a hill. Indeed, published work indicates that about ninety per cent of bodies are discovered downhill.

MASS GRAVES

A murder is a horrific thing; a multiple murder more horrific still. A mass murder, involving hundreds or thousands of victims, must be the most horrifying and repugnant thing human beings can do. The genocide of the Jews by the Nazis; the Bosnian massacres of recent years; the events in Kosova, Cambodia, and Rwanda – these must be among the most dreadful things imaginable.

No experience can compare with a visit to the mass grave of people massacred because they had the "wrong" racial or cultural background. For me, it is certainly not an experience I would wish to repeat. No-one, not even the most detached forensic scientist, can remain unmoved by such a thing. This does not mean that such feelings should, or do, interfere with the work of a forensic investigator; but it does mean that one's view of the world is deeply affected. Such matters may appear to lie outside the remit of this book, but, in my view, they do not.

Anyone wishing to take up a career in forensic science should consider seriously how they might react to horrors of this kind. Seeing horrific things is part and parcel of forensic investigation; and it is no use at all being highly-skilled at the scientific aspects alone, if one cannot deal emotionally with the sordid realities to which such an investigator is exposed.

In Rwanda, behind the church at Kibuye on the shore of lake Kivu, a large number of bodies were buried after their massacre. No-one knew exactly how many bodies were there; estimates ranged from a few hundred to two thousand. Six months after the massacre, the grave was opened. In situations like this, the investigators must try to arrive at an estimate of the least number of people interred; in other words, the smallest number of people that had definitely been killed. The difficulty over what may appear a simple matter of counting the bodies lies in the fact that the bodies are usually decomposed; parts are missing and the bones are broken into small fragments and mixed up together in such a way that it is impossible, in most cases, to determine any discrete bodies.

A particular bone, often the bones of the skull collectively, is used as a marker. If one can find a definite number of separately recognizable skulls, then the number of bodies in the grave must be at least that number, possibly more. In Kibuye, the figure arrived at was 460. Other grim facts emerged; from a study of the way the wounds were inflicted, it was concluded that at least sixty-five per cent were bludgeoned to death – about three hundred people. Another conclusion was that forty-four per cent were children – about 202 children. Thus, the detached methods of forensic science reveal the extent of human barbarity.

The bones of the dead, if left exposed above ground, will scatter over the surface, as a result of the activity of wild animals. The study of such scattering is called taphonomy, and it is a subject that, until very recently, was the province of palaeontologists interested in the patterns of distribution of fossil bones and what such patterns can reveal about the past. The extent to which skeletons are disarticulated and scattered is often correlated with time since death. Although such estimates are necessarily very general or vague, forensic science now uses such techniques to arrive at a general reconstruction, before the reconstruction can be fine-tuned using more precise methods. But it is not only the activity of wild animals that affects the positions of bones; the results of human activity can often explain what happened before death.

The Bosnian countryside is littered with such bones. Reconstructing events prior to the death of the individuals concerned is all-too-often an easy matter. Near Kravica, the bones of a youth, who could not have been more than seventeen, lie on a stretcher, the long bones jutting out of the trouser leg. The bones of older men lie beside the stretcher. All were shot as they were trying to take the injured boy to safety.

Down the hill, toward some woods, there are more skeletons. People had been fleeing to the woods for safety and were shot in the back as they fled. Some just made it into the woods, such as a young man in his twenties, his shattered ribs telling of the bust of machine-gun fire that killed him.

It is estimated that five thousand people died in the massacre at Srebrenica; and that about eight thousand are still missing, probably dead. John Gearns, an American forensic scientist investigating the Bosnian massacres, ordered the placing of plastic sheeting, weighed down with soil, over all the burial sites in the area, since the bodies of the dead had attracted the attentions of scavenging wild animals.

Dr William Haglund, an American forensic anthropologist, investigating mass graves near Nova Kasaba, used a T-shaped rod, one inch in diameter, to find bodies. Pushing it into the soil, then taking it out and sniffing it, he tried to detect the odour of human decay. When such an odour is detected, the digging begins. Gentle and courteous, Dr Haglund would refer to the victims with great respect: "We have gentleman no. 1 right here, in a sitting position, sort of on top of other people. Here, you can see gentleman no. 2 is lying on his back. His head is in a downward position. His hands are beside him, like this [he gestures] and bound. Gentleman no. 3 is also lying on his back and his hands are behind his back."

Many others had their hands tied behind their backs, to be thrown "ignobly in a hole" in the words of Dr Haglund. Many had their skulls shattered by high velocity bullets. The shell casings found among tightly packed bodies suggest that some of the wounded were shot as they lay among the dead. The way the bodies were lying suggested to Dr Haglund that the "scene is consistent with these people having been lined up along the side of the road and shot". It was dangerous work; mines had to be cleared and the excavation had to be conducted under the protection of NATO guards. It was also harrowing work. "You use one part heart and whole part brain", said the Chilean forensic scientist, David Del Pino.

Dr Haglund and his colleague, Dr Clyde Snow, have also investigated mass graves near Vukovar, Croatia. At the Ovcara Farm outside the town, a detailed map was made of every object found – every bone, every bullet was documented. Snow took down the descriptions of missing people from friends and relations and entered them into a computer database for later comparison with reconstructed individuals. It is slow and painstaking work; and it can take many years for the results to reach scientific journals.*

Meanwhile, the work continues. It can never end. "This book will never be completely closed," said a representative of the Red Cross. Dr Haglund agrees, and explains the purpose of his work:

"The real reason you're dealing with the dead is because of the living. You do it because you're concerned with other people and concerned about resolving the questions that leave their lives in limbo. They're missing someone, and they can't continue with their lives. They have this vacuum. and they fill up the vacuum with hope. That person might still be alive – in forced labour camps, incarcerated somewhere, working in mines. With my information, I can put an end to some of the questions. I try not to be a spoiler of dreams. Maybe just a spoiler of nightmares."

Such is the spirit of place. Forensic investigators must adapt – scientifically and emotionally – to the special requirements of each scene of tragedy as though it were unique. No two murders are the same.

MEDIUMS AND SPIRITUALISTS

There remains to discuss one method by which missing people and buried bodies have reputedly been discovered, although it is a technique that has been much derided, perhaps rightly so. However, the forensic investigator must not scoff too soon at subjects that may only be in their infancy and which may have great potential for the future. Let us keep an open mind. The method in question is the work of mediums and spiritualists.

"How can a scientist take such things seriously?" you may ask. Is this a respectable book about forensic science, or is it a collection of crankish notions? Bear with me and you will hear the answer.

Agatha Christie, the famous writer of whodunits, disappeared from her home on Friday December 3, 1926, after a row with her husband, Colonel Archibald Christie. The story has been told many times and a speculative motion-picture was made about what she did during the days of her disappearance. I do not propose to go into this story in detail, simply to mention one aspect of it.

As the days passed, with no news of what became of Agatha Christie, the Chief Constable of the Surrey Police sought the assistance of Sir Arthur Conan Doyle. Conan Doyle agreed to assist in this case of a missing fellow author, but he did not employ the methods of Sherlock Holmes. What he did was to ask the police for one of Mrs Christie's gloves. This he took to a spiritualist friend, one Horace Leaf, telling him nothing at all about the owner of the item. Leaf immediately came up with the name "Agatha". He then said: "There is trouble connected with this article. The person who owns it is half-dazed and half-purposeful. She is not dead as many think. She is alive. You will hear of her, I think, next Wednesday."

On Wednesday morning the newspapers reported that Agatha Christie had been found alive and well.

What should one make of this? Was it a bizarre coincidence? Perhaps. Did Horace Leaf assume that the glove belonged to the famous writer and decide to hazard a guess that proved correct? Quite possibly. Could the fact that the inquiry came from Conan Doyle, a fellow author, and the fact that it concerned the police in Surrey, where Agatha lived, have helped the medium to guess correctly? This, too, is possible. What, then, is my point?

Simply this: there is much that we do not understand about the world we live in. There is evidence to show that telepathy is a real phenomenon; indeed, in many parts of the world it is considered the norm; and people in, say Africa and South America, have expressed disbelief that western Europeans do not have this faculty.

Am I saying, then, that I believe in spirituality and telepathy and mediums and all the other things that are usually regarded as mumbo-jumbo? Well, yes and no; and I am not being evasive. An observed phenomenon is an observed phenomenon – one cannot simply dismiss it because one links it in one's mind with the supernatural or the occult. A phenomenon may be given a supernatural explanation, but that does not mean that it is supernatural, nor does it mean that the phenomenon itself does not happen. People used to believe that volcanoes erupted because a subterranean dragon's anger was aroused. Nowadays we do not believe such things, but the phenomenon of an erupting volcano has not disappeared – we simply have a better explanation for it, an explanation that seems logical and which does not strain our credibility. Calling a phenomenon supernatural does not make it supernatural. What we really should be dismissing is the explanation, not the phenomenon itself.

This confusion between words and reality is commoner, even among scientists, than is generally realized. For example, most scientists dismiss the notion of faith-healing, a phenomenon for which there is a certain amount of evidence. Some people, simply by placing their hands close to someone's body, are able to generate a localized sensation of warmth in that person. Interestingly, a thermometer placed on the warm region will not register a rise in temperature. Clearly, something is stimulating the nerve endings that normally register an increase in temperature, but it is not an actual rise in external temperature that is making them respond in this way. These are facts; what they mean is another matter.

I was once discussing faith-healing with a fellow scientist, who became somewhat impatient. He told me that I would not do my reputation any good by taking such ideas seriously; for his part, nothing would induce him to believe in such nonsense. And yet the facts were clear enough. As I told my colleague, the reason he did not accept the whole phenomenon was that it was called faith-healing. But this is the name given to the phenomenon, nothing more. If it had been given another name, especially a very "scientific" sounding one, like Simulated Temperature Increase Phenomenon (STIP), he would have had no difficulty at all in discussing the subject. He was put off by the name "faith-healing", implying as it does a supernatural and mysterious process. So-called faith-healing is no doubt a perfectly and rationally explicable phenomenon, but since we do not yet know its cause and because of the name given to it (which suggests that the "cause" is known) is seen to be objectionable, the baby went out with the bath-water. And so it is with telepathy; if we can begin by accepting that the phenomenon exists and not endow it with any mystical explanations that appeal more to the emotions than to reason, we might be able to investigate it rationally. It may be a rewarding study, or it may not; but dismissing out of hand, because others pretended to have already discerned its "cause", is not good science.

---

*It was eventually concluded that there were some 200 victims in the Ovcara Farm grave. To date DNA identification has been made in over 90 per cent of cases.

CHAPTER THREE

TIME WILL TELL

There's a time for all things.
William Shakespeare
A Comedy of Errors
Act II. Scene 2

It was an unseasonably hot day in September. The murdered body of a man was lying at the base of a tree in the wood, surrounded by policemen, scene-of-crime officers, photographers and forensic scientists from the Home Office. The smell from the decomposing body was overpowering, even there in that outdoor place. The pathologist and I were standing somewhat apart from the rest, discussing a question that was central to the case. When?

When was the murder committed? This question is one of the most frequently asked during a murder investigation. Whodunits and mystery novels almost always have an answer, which is usually provided by the doctor at the scene – "2.45 am last Tuesday," he might say. So widespread is the perception that estimating the time of death is a simple, routine matter, that it might come as a shock to learn that, of all forensic science questions, it is probably the most difficult to answer. In fact, when the first few days after death have passed, conventional forensic pathology has little to contribute to this question.

This does not mean that there are no methods available for dating the time of death. In fact, there is a whole armoury of techniques, but none of them can be applied routinely to any and all cases, except, perhaps, for some of the conventional pathology methods.

What, then, are the methods used? In principle, anything that changes with time can be used as a clock, as long as we understand how that clock works. The human body normally loses heat by radiation, convection and evaporation, but at the same time it generates heat as a by-product of muscle activity, chemical breakdown of ingested foodstuffs and so on, with the result that a healthy person maintains an average core body temperature of around 37.5 degrees Celsius (99.5° Fahrenheit). Almost as soon as someone dies, their ability to regulate their core temperature ceases. In cool and temperate climates there is an initial period – the temperature plateau – during which their body temperature is maintained, but after this their body will begin to cool down, a process known as algor mortis. The relationship between the rate of this cooling and the passage of time is reasonably well understood, although practical problems do arise and there are some caveats. Under what may be called "room temperature", after the initial plateau, which is generally thought to last for between half an hour up to three hours, a dead body will lose heat at a linear rate until it reaches the temperature of its environment. By the time twenty-four hours have elapsed, the body temperature will be the same as that of the surrounding air. The body usually feels cold to the touch about twelve hours after death.

The above describes a fairly "typical" situation, but conditions are not always typical. If the deceased had died of asphyxiation or a cerebral haemorrhage, the initial temperature may actually be raised. The late Professor Keith Simpson once measured an initial temperature of 43°C, several degrees higher than the normal temperature of the living body of 37°C. A naked body will lose heat faster than a clothed body; and one submerged in water will cool more rapidly than one lying on land. In general, larger people lose heat more slowly than ones of slighter build, although this is not always the case.

It is important to take the body temperature – ideally taken at the rectum – at the scene, i.e. before removing the body to the mortuary. In some cases, such as those of suspected sexual assault where trace evidence might be damaged by taking a rectal temperature measurement, the intra-abdominal temperature can be measured by means of an abdominal stab. Nowadays, digital probe thermometers are commonly used. The air temperature is also taken, so that the relationship between the two can be established. Ideally, two sets of temperatures should be taken about one hour apart, since this would make the relationship between body and air temperatures clearer. Various, somewhat rough-and-ready, guides to time of death based on temperature have been developed over the years. One commonly used formula was:

(normal body temperature in Fahrenheit minus measured body temperature) divided by 1.5 = number of hours since death

Because so many factors can have an influence on a body's rate of cooling, such formulae are now regarded as being too simplistic. Instead nomograms (two-dimensional graphical calculators), such as that developed by Henssge which uses a more complicated formula to link the variables, are more commonly used. Using Henssge's nomogram, an estimate is made of the weight of the body, the rectal and air temperatures are taken and an estimate of the number of hours since death can be obtained.

Cooling rate is the classic method of determining time of death in the early stages, but, as we have seen, this method can only be used during a very short period after death. Another change often used is hypostasis, also known as post mortem lividity, which involves the appearance of reddish-purplish coloration in the skin. We discussed this phenomenon briefly in the last chapter. Hypostasis takes place because blood ceases to circulate after death and, obeying the law of gravity, it sinks to those parts of the body that are lowermost, the red blood cells sinking first. In other words, a body lying face down will exhibit lividity in the abdomen, whereas one lying on its back will show livid patches on the back and the backs of the thighs. Those parts of the body lying compressed against the ground will appear white, since the blood vessels will not be able to fill up. Lividity appears within two hours after death and becomes marked after about four or five hours later, when the initial patches of lividity fuse into a more continuous discoloured area. In the early stages the pressure of the hand will whiten the skin, since the blood will be forced away, but later, when the blood has coagulated, the purplish stain will not disappear under pressure. Like post mortem cooling, hypostasis is a very limited guide to time of death.

Rigor mortis is, perhaps, the best known post mortem change, yet it, too, is highly unreliable as a means of establishing when a person died. Stiffening of the muscles, caused by the accumulation of certain salts in the muscle fibres, appears about twelve hours after death. The face muscles stiffen first, then the shoulders and arms and, finally, the legs. Rigor is at its most pronounced during the next twelve hours, after which it takes another twelve hours to disappear completely, the body losing its rigidity from the face first, followed by the other muscles in the same order as rigidity took place.

A particular form of muscular rigidity, sometimes incorrectly interpreted as rigor, is the phenomenon known as cadaveric spasm, an event much used by writers of whodunits. This occurs at the very moment of death and is expressed as a sudden tightening of muscles, especially of the hand, in which an object, such as a weapon or a piece of paper, may be grasped. This is a rare occurrence and it is not understood why it happens, although when it does happen it may shed light on the course of events, much as the mystery writers have shown.

The decomposition of the body can give some clues about time of death. About two days after death a "marbling" pattern appears at the neck and shoulders and runs down the arms and sides of the abdomen. At first, this appears red in colour, turning green later. These changes are caused by the invasion of bacteria from the tissues into the blood veins. At this time the body becomes bloated, filling with gases generated by bacterial metabolism, and the features become unrecognizable. The internal organs are now decomposing and the order in which they decompose may give some pointers to time of death.

At first the stomach and intestines, as well as the heart and the blood in the liver, begin to decompose. Next the lungs and the liver itself break down. This is followed by the brain and the spinal cord. Next, the kidneys, bladder and testis begin to decay, followed by the general body musculature. The uterus and the prostate gland break down last. These, then, are the main gross changes in the body that could be used, in a broad brush-stroke kind of way, to narrow down the time of death. Let us now look at some more specialized techniques.

MAGGOTS

When a dead body is exposed in nature it soon becomes attractive to other forms of life. Carnivorous animals will come to eat, finding an easy meal. Insects will arrive to lay their eggs. Crows, rats and mice may come to peck or nibble at the remains. All this may sound macabre, but it is a fact of life; moreover, it is a fact that can enable us to estimate the time of death.

Flies of one kind or another are usually the first to arrive at a dead body. Such a body, exposed above ground during the warmer months of the year, will attract bluebottles and greenbottles. The flies will lay their eggs in the natural body orifices (ears, nose, eyes, etc.) and the maggots that eventually hatch will feed on the muscle tissues of the remains. They will grow, shed their skins, feed a bit more, shed their skins again, then feed and grow until they have extracted enough nutrients from the body to enable them to pupate. The last maggot stage will then leave the body, burrow in the soil (or crawl under an object, such as a carpet, if indoors) and pupate. Its outer skin becomes hard and it darkens and contracts, forming the pupal case, or puparium. Eventually, the adult fly will emerge from its case and the cycle will start all over again.

As it happens, maggots of various kinds are very intensively studied creatures, not only because of their forensic interest, but also for their relevance to medical and veterinary problems and for the simple fact that, being easy to keep in laboratory colonies, they are used for studies in development and genetics. This means that we know a great deal about them; and this knowledge is put to good use in forensic investigations.

The basic question that a maggot can help us to answer is this: "What is the minimum time that has elapsed since death?" In other words, what is the time after which death could not have occurred? For example, if we can estimate the age of a maggot as being, say, five days old, then we can say that death could not have happened later than five days before discovery of the body. Knowing the age of the maggot cannot tell us the actual date of death, since we do not know when the fly arrived at the body, only the minimum time that must have passed since the crime was committed. The more tightly we can make the minimum time of death estimation, the more firmly we can place hand on heart and say we are confident of our conclusion.

However, if we have reason to believe that the body had been exposed to fly activity almost as soon as it was dead, then we can say that the minimum time since death is the actual time of death. The ability of flies to detect the odour of dead body, even at the very earliest stages, is so well developed that they will reach it well within an hour after death.

How does one determine the age of a maggot? Like all living things, with the exception of birds and mammals, the rate of development of maggots is determined by the external temperature, since they cannot control their body temperature by internal physiological mechanisms, as we can. Therefore, in order to estimate a maggot's age one has to have a very good idea of the temperatures that prevailed during the course of its development. So, one needs to take the temperature of the maggots in the body, which can be many degrees above the surrounding temperature, adding another complication to the estimation of time of death based on body cooling. In some cases, the air temperature may be of the order of 20°C, whereas the maggot-infested body may have a temperature as high as 40°C.

While it is an easy matter to measure the temperature of the maggots and the surrounding air at the scene, it is much more difficult to estimate the temperatures that prevailed in the days before the body was discovered. Consequently, one has to find another source of information, which is provided by the meteorological offices. Unfortunately, this is only the beginning of the solution.

As we have seen, it is important to take at least two sets of temperature measurements at the scene, in order to establish the relationship between them. Despite the fact that we know a great deal about the way surrounding temperatures affect the temperature within a maggot-infested body, there are always surprises in store, since local conditions may not reflect the conditions under which laboratory studies were conducted. Another complication is that weather stations offer only a general guide to temperatures, since meteorologists are concerned mainly with the "big" weather picture and avoid the confusing details of the climate near the ground. This is why weather stations are always situated in open countryside and why the grass around them is mown very short.

Unfortunately, in forensic work of this kind it is the temperature near the ground – the microclimatic temperatures – that are of interest. Fortunately, however, temperatures of the air far above the ground can help us to reconstruct the temperatures near the ground, or what meteorologists like to call the boundary layer. Having reconstructed these, one can then reconstruct the temperatures within the maggot-infested body.

Now that the temperature regime has been reconstructed, it is necessary to dissect the maggot and determine its stage of development. A maggot's stage and its age are not the same thing; in fact, they are very different things. The stage is the degree of its development; the age is how long it has been alive. At low temperatures maggots develop more slowly than they do at higher temperatures (generally speaking); in other words, they live longer at low temperatures. So, one maggot may be at a more advanced stage of development than another maggot, which latter is, nevertheless, older than the former. It is the stage and the temperature, taken together, that will reveal the age of the maggot.

ESTABLISHING THE TIME OF DEATH

Time of death is such a basic question in murder investigations that it is not at all easy to pick out a particularly good example to show how it was used to solve a crime. However, two cases are worth discussing. First, a case from my experience – one of the most unpleasant cases with which I was involved. It is the case of Sidney Cooke and his paedophile gang.

The body of Jason Swift, a 14-year-old boy, was discovered by a gamekeeper in a thicket inside a woodland early in December, 1985. The body had some maggots in it and, on the basis of these, the police asked for a time of death estimation. The weather had been very cold, dropping to well below freezing during the days before the body was found. On the day of discovery, it rained and the temperatures rose to above freezing.

The problem here was how any maggots could have been in the body at all, since the temperatures prevailing between the time of the boy's disappearance to the time his body was discovered would have prevented any flies from being active and laying eggs. Therefore, the body could not have been lying out of doors when the flies arrived to lay their eggs on it. The maggots, at their second stage of development, were too old to have been derived from eggs laid on the day the temperatures rose and the body was discovered, so they must have been developing indoors. I concluded that death must have occurred at least two days prior to discovery and that the body must have lain indoors for some time. These facts, slight though they are, were corroborated by other evidence and the minimum time since death estimation turned out to be the actual time of death. The murder itself took place in a car.

Our next example is a particularly good one, since it demonstrates, not only how much depends on the time of death, but also because it shows what can be achieved by a forensic investigator with imagination and flair. The man in question was the eminent forensic pathologist, Sir Sydney Smith.

On October 9, 1951, at around 5 pm, the Ceylonese cricketer, Sathasivam, was arrested and charged with the murder of his wife in a pleasant suburb of Colombo. Some time after 3 pm that day, Mrs Sathasivam had been found strangled to death in the garage of their house.

The facts were very simple: Mr and Mrs Sathasivam were getting a divorce and, although Mr Sathasivam had been staying with a friend for some time, he returned home on October 8 and stayed the night. On the day of the murder, the ayah – or children's nurse – had breakfast with the couple's two older children, who left for school at 8.15 am. At that time, Mrs Sathasivam was in the younger children's room, reading her letters and the morning paper. Mr Sathasivam was still asleep. The only other person in the house at that time was the cook, William, a youth of eighteen.

When the ayah and the older children returned home at about 3 pm, Mrs Sathasivam was dead, lying on her back on the garage floor. The smaller children, unaware that anything was amiss, were playing in the kitchen. Neither Mr Sathasivam nor William were in the house.

At 6 pm the pathologist, Professor de Saram, examined the body on location in the garage. The following day he carried out a full post-mortem examination. He reported on his findings and gave his opinion that death had occurred some time between 10.00 am and 11.30 am.

Now, Mr Sathasivam left the house that day at 10.35 am by taxi. That he did so was proven beyond doubt. He could also account for his movements for the rest of the day. In other words, if he had committed the murder, he would have had to have done it before 10.35 am – otherwise he was innocent. The evidence of Professor de Saram left the matter of guilt or innocence open.

Ten days later, William, who had disappeared, was found and questioned. In his statement, he said that Mr Sathasivam had murdered his wife and that he had assisted in the crime, because his master ordered him to do so. Moreover, he said that the murder was committed before 9 am and that after the murder, Mr Sathasivam removed Mrs Sathasivam's jewellery, including a gold necklace known as the Thali Kodi, a sacred possession of Hindu women, who would never remove it as long as their husband was alive. William said that Sathasivam gave him the jewellery, plus three rupees, as a reward for his services.

William then went to a jeweller's shop and sold the jewellery – this was at 9.30 am. The two men at the jeweller's confirmed the time. However, the manager of the taxi company testified that a woman telephoned him at about 10.30 and asked him to send a cab to her house. Moreover, he said that he recognized Mrs Sathasivam's voice. Also, the driver of the cab said that he saw Mrs Sathasivam at the door when he arrived to collect Mr Sathasivam. Another witness was Mrs Sathasivam's lawyer, he said that he had had a telephone conversation with her some time between 10.25 am and 12 noon.

The testimony of the jewellers had to be weighed against the evidence of the three other witnesses. The jewellers were known to be dealers in stolen goods and were generally regarded as unreliable witnesses; the other three were respectable people, who had no reason to lie.

The prosecution accepted William's evidence, although it conflicted with de Saram's conclusions. So, the situation arose where the prosecution did not accept the testimony of its own medical witness, but did accept the evidence of the cook, who, by his own admission, was at least an accessory to murder, as well as being the recipient of stolen goods.

William's story benefited from a certain realism – a realism that could have easily derived from the fact that he was the murderer, not an accessory. He said that the murder was committed in the bedroom upstairs. Mr Sathasivam allegedly took his wife by the hair with his left hand and grasped her by the throat with the right hand, then pulled her on to the floor. She fell, face upward. Sathasivam, still holding his wife by the throat, ordered William to hold her legs, which he did. Then Sathasivam stood up, placed his shod foot on his wife's throat and pressed down hard; William said that he heard the voice-box crack. Finally, they carried the body down stairs, Sathasivam holding his wife under the armpits, with William holding her feet.

In view of the conflict between William's evidence and de Saram's conclusions, Sathasivam's lawyers decided to contact Sir Sydney Smith of the University of Edinburgh for advice. Smith was supplied with all the relevant documentation. The evidence that the woman had been strangled was clear enough; the voice-box was, indeed, broken. However, it seemed unlikely to Smith that William could have heard it break, although he could easily see how someone pressing down on the throat with a bare foot would feel it break. William went around in bare feet.

The medical report showed that there was an abrasion on Mrs Sathasivam's jaw and neck, presumably where the trampling had been done. There were fibres and a fine dark powder on the neck, identical to those found on the floor of the kitchen, according to the findings of the government analyst. To Smith this evidence suggested that the fibres and powder came from William's bare foot as it curved over the jaw-bone.

There were two other bruises; one on the side of the head and the other on the back, between the shoulder blades. The first was clearly caused by a blow from a blunt object, although the prosecution asserted that it was caused by bumping the victim on the floor. They further alleged that the second bruise was caused in the same way, but, again, Smith disagreed and said that the bruise was probably made by some projection, such as a nail, on a door or wall.

William had stated that the victim was killed while lying on her back on the bedroom floor, but another piece of evidence told against this. The victim of strangulation usually passes urine just before they die; and the urine stains on the deceased's petticoat ran almost down to the hem. This would have been unlikely to happen if Mrs Sathasivam had been lying down at the time of the murder.

The temperature of the body was 93.2°F at 7 pm on the day of the murder – a loss of only 5.2°F. In view of the fact that the body was small, the clothing was thin and that the deceased lay on a cement floor – all factors that would increase the rate of cooling – it seemed odd that the body lost only a little over 5°F in ten hours. Experiments showed that, under the same conditions as the murder, bodies lost 5.2°F in seven hours. However, the conditions were not exactly the same, since the bodies of executed men on which the experiments were carried out were much larger than the body of Mrs Sathasivam; in other words, less than seven hours would have been needed to reach a temperature of 93.2°F. This indicated that death must have taken place later than 10.35 am.

Rigor mortis was almost complete by 6 pm, when Professor de Saram examined the body. For a body to have reached that stage under the prevailing conditions would take between six and eight hours, placing the time of death somewhere between 10 am and 12 noon, although, in Smith's opinion it was more likely to have been closer to noon than to 10 am. In any case, William had said that death had occurred before 9 am.

William had said that he and his master had carried the murdered woman to the kitchen and then to the garage and that the body never touched the floor. However, as she, like William, walked about in bare feet, there was a layer of fibre and dust, identical to those in the kitchen, on her feet. This seemed to suggest that she walked into the kitchen and did not leave it alive.

Nor was this all. There were drag marks on the kitchen floor, a fact that suggested that the victim was dragged from there to the garage. Mrs Sathasivam had been wearing an ear-stud that day, but the pearl that had been attached to it was missing. It was discovered twelve days after the crime, lying about a foot from where the body lay. There was bruise behind the left ear, which must have been caused by the pressure of the stud during the assault; as the bruise was caused while the victim was still alive, it suggested that the murder took place in the kitchen or the garage, which was adjacent to it.

Most interestingly, William's face and arm were scratched, whereas Mr Sathasivam was not. The cook said that he received these scratches when he obeyed his master's order to hold the woman's legs, but it was much more likely that he was scratched while struggling with the woman himself. Moreover, there was seminal fluid on the woman's sari, which was torn in several places. In order to reconstruct the crime, Smith instructed himself on the way a sari is wrapped around the body and arrived at the conclusion that the tears coincided with the wounds on the back.

In May, 1952, Sir Sydney Smith arrived in Colombo and went straight from the airport to the scene of the crime. There he found on the wall between the kitchen and the garage a staple that was used to hook the door back when it was open, the very thing he predicted he would find when he was several thousand miles away in Edinburgh. Sir Sydney Smith gave evidence at the trial and Sathasivam was found not guilty. William was never tried.

WHY DETERMINE THE TIME OF DEATH?

It is not only the unlawful killing of human beings that can result in prosecution; the unlawful killing of animals, domestic and wild, may also arouse the interest of the police. Dr Robert Stebbings, an authority on the biology of bats, once telephoned me to ask for help in a case he had been consulted upon. A man had telephoned the Nature Conservancy Council (NCC) and told them that he had bats in his attic and that he wished to get rid of them. Under the law it is illegal to kill bats; the proper course of action is to ask the NCC to remove them. In this case the NCC told the man that they would arrive at his house the following day, but they did not. They did appear the day after that, but, when they examined the attic, they found that all the bats were dead. When the NCC officials took the man to task for having killed the bats, he protested that, since they had not turned up the day before, he decided to kill the bats that morning.

The bats' bodies were infested with maggots. Moreover, they were quite old maggots, being at least six days old; and it was clear that the man had killed the bats before he even telephoned the NCC two days earlier. Why he did so, after killing the bats, is a mystery, but there was no doubt that he had killed them at least six days before the NCC officials arrived. He was successfully prosecuted.

Another case, brought to me by my colleague Dr Tim Benton of Stirling University, concerned a badger that was found in a wire snare in Scotland. The badger, although still alive, was heavily infested with maggots, a condition known as myiasis. Certain fly species can cause this condition by laying eggs on the living animal. The police wanted to know whether unnecessary suffering was caused to the badger by the negligent behaviour of the person who laid the trap. Under the law, such traps should be inspected every day. The maggots in the badger's body were almost fully-grown, so it was clear that the trap had not been inspected for up to a week. The animal was too badly injured and had to be out down.

The application of methods of time-of-death estimation in cases of killings of endangered species is increasing, especially in North America. As the protection of rare animals is now a legal matter, many issues of wildlife conservation are becoming part of forensic science investigations.

As time passes and decomposition progresses, different insect species will be attracted to the body at different times. This succession of fauna is impossible to predict, but is quite useful in the reconstruction of events after the event.

I am often asked how this can be – how can an unpredictable phenomenon be used as a forensic tool? The answer is that forensic science, unlike much other science, is not concerned with predicting the future, but reconstructing the past. If, at the crime scene, one finds a cigarette-end and a used match, one is entitled to conclude that a smoker had been there. One could not possibly have predicted such a thing beforehand, but finding the evidence enabled us to arrive at that conclusion. The same is true of the insects, or their remains, found on a body. If we should find, say, the pupal cases of a fly that is active only in the spring, then that fact will help us to narrow down the time of death.

This reasoning, of course, applies to much forensic thinking. If, in December, one finds some shrivelled seeds of oilseed rape in the jacket pocket of a murder victim, one would naturally reflect on what could have happened in May, when the plant was growing and setting seed. If the deceased had been wearing a shirt of a kind that was first manufactured only six months ago, then one is entitled to conclude that death occurred after that date. Forensic scientists should not be too fussy about the kind of evidence available; they must use to maximum effect whatever comes to hand. Evidence involving the use of high-tech apparatus is not always the most revealing; very often the simplest of simple clues can be the most revealing.

This brings me to another point: the forensic scientist must look for clues wherever they may be found. In an earlier chapter I emphasized that it is the way forensic scientists think that matters, not so much the techniques at their disposal. A microscope and a computer no more make a forensic scientist than a brush and paints make an artist.

With these thoughts in mind, let us pause for a moment and ask why it is so necessary to determine the time of death. The answer is that the time of death will tell us so many other things. Could Bloggs have been the murderer? No, he was elsewhere during the critical minimum time since death period. Could Jones have done it? Possibly, since he was in the vicinity at the time; so he remains under investigation. So, the time of death can either include some suspects in, or exclude them from, an investigation. The time of death can tell us other things, even how death occurred. Let us look at an example of how this can be.

A VERY FAMOUS CASE

Our example is concerned with a crime that was committed almost two thousand years ago, yet it is a crime that has affected, in one way or another, everyone who has lived since then. It is the crucifixion of Jesus Christ. What is the truth? What really happened? Can we really tell what happened, at this great remove in time? I believe we can.

It was Pontius Pilate who asked, "What is Truth?", so it seems appropriate that he be allowed to guide us toward the truth in this matter. Let us look dispassionately at what happened. On the whole, the accounts of the four Gospels agree about what happened. After his arrest and torture, Jesus was sent to Pilate and, subsequently, to the High Priest, Caiaphas, for interrogation. He was then sentenced to death, the mob having chosen Barabbas for release, rather than Jesus, who was then crucified.

The events up to this point are quite straightforward, but a puzzling event takes place while Jesus is on the cross. It is this: he dies, quickly. This is most unusual, since crucified people normally lingered on for days in a slow, truly agonizing death. It is true that Jesus was much weakened by a severe whipping before his crucifixion, but so were other condemned people. Why did Jesus die so quickly?

Consider what happened next. This is what the Gospel according to St Mark (Chapter 15, verses 43-45) says:

"Joseph of Arimathea, an honourable counsellor, who also waited for the kingdom of God, came, and went in boldly unto Pilate, and craved the body of Jesus. And Pilate marvelled if he were already dead: and calling unto him the centurion, he asked him whether he had been any while dead. And when he knew it of the centurion, he gave the body to Joseph."

So, Pilate was surprised that Jesus died so soon after he was crucified. His reaction is one of scepticism, for he knows that his crucified victims normally took several days to die and he asks a centurion to check Joseph's account and report back to him. The centurion goes on his errand and returns with his report: yes, Jesus is dead. Pilate, perhaps still somewhat astonished, then agrees to let Joseph have the body.

Although all the Gospels mention Joseph of Arimathea and his interview with Pontius Pilate, only St Mark records Pilate's surprise at the early death. But there is another source of evidence that tells us that Jesus's early death was unexpected by the Romans. This is supplied by St John (Chapter 19, verses 31-33):

"The Jews therefore, because it was the preparation, that the bodies should not remain upon the cross on the Sabbath day (for that Sabbath day was an high day), besought Pilate that their legs be broken, and that they might be taken away. Then came the soldiers, and brake the legs of the first, and of the other which was crucified with him. But when they came to Jesus, and saw that he was dead already, they brake not his legs."

The purpose of breaking the victims' legs was so that they would die quickly; they then could be removed before the holy day. When the soldiers arrived, they broke the legs of the two criminals (incidentally, there is nothing to suggest that they were specifically thieves, as is often stated) who were crucified with Jesus, but when they came to Jesus, they discovered that he was already dead. Three men crucified at the same time, but only one of them was dead at a time when he would have been expected still to be alive. Hence the need to speed things up and break the legs.

The crucifixion of Jesus is such a familiar event that its true horror may not always be apparent. Imagine the pain and humiliation of it; a man stripped naked, horribly whipped, is paraded through the streets and made to carry a wooden cross. He is then hung naked on a gibbet and left to die, while others jeer at him and mock him in his agony. Other events, equally unpleasant, tend to be overlooked or taken for granted as part of the well-known story. After the soldiers arrived and saw that Jesus was dead, what happened next was this (St John: Chapter 19, verse 34):

"But one of the soldiers with a spear pierced his side, and forthwith came there out blood and water."

This unnecessary and barbaric mutilation of a dead body has not, to my knowledge, been explained. It seems reasonable to speculate that the Roman soldier, having come on his leg-breaking mission with some glee, felt cheated that one of his intended victims was dead. Why else would he do such a vile thing as stab a dead body with a spear? I submit that this vicious deed was carried out because the soldier found the early death totally unexpected and saw fit to vent his spleen on the victim who cheated him. I cannot think of a better explanation.

Let us now turn back to the last horrific moments before Jesus's death. In agony, he was praying to God: "*Eli, Eli, lima sabachthani.*" ("My God, my God, why hast thou forsaken me?") The gathered crowd believe that he was calling upon Elias (or Elijah) and a man, carrying a sponge soaked in sour wine or vinegar at the end of a long cane, rushed up and pushed the sponge at the face of Jesus. Jesus then gave a loud cry, and died. This is what St Mark says (Chapter 15, verses 35-37):

> *"And some of them that stood by, when they heard it, said Behold he calleth Elias. And one ran and filled a sponge full of vinegar and put it on a reed, and gave him to drink, saying, let alone; let us see whether Elias will come to take him down. And Jesus cried with a loud voice, and gave up the ghost."*

The Gospels of St Matthew and St John tell essentially the same story, although it is absent from St Luke's. St John does not say that Jesus cried out after receiving the vinegar.

What is the significance of all this? When a man is crucified, his arms are stretched along the horizontal bar of the cross, making breathing difficult. The breath comes in short inhalations and exhalations, since it is difficult and uncomfortable to take a deep breath in such a position. When the vinegar-soaked sponge was put against his face, Jesus, because of the shock of the pungent smell, must have inhaled deeply, which is the natural reaction under such circumstances. But, having done so, it would have been very difficult for him to exhale once more. The muscular anatomy of the chest makes this effectively impossible. A painful death from asphyxiation would follow soon after. The last, loud cry of Jesus was probably the sudden, painful inhalation.

This is probably why Jesus died so quickly on the cross. Other victims were, presumably, left alone to die slowly, since most such people would have had the sympathy of the Jewish population, all crucifixions or sentences of death being carried out by the Romans, who alone could pass a sentence of death. Jesus, however, was resented as the man who claimed to be the "King of the Jews", and was tormented as he died. The incident of the sponge was the last straw.

What does all this mean? Why should such a seemingly trivial event as the sponge incident help us to understand whether or not the whole episode really happened? The reason is this. The Gospels are essentially religious writings. They were written to make a theological point and to spread the news of Jesus's life, death and resurrection. They were not written as historical treatises. Those who wrote them were not interested in matters that were not central to the argument. Therefore, they would not have taken the trouble to add irrelevant information, in other words, information that did not help them make their point, unless the events had actually been witnessed. The point here is that nothing turns on the early death of Jesus; it proves nothing and the event is not used by the authors of the Gospels to prove anything. It simply does not matter, theologically. It happened and was recorded, although nobody knew, or claimed to know, why or how it happened. To put it bluntly, Jesus's early death is such strong evidence that the crucifixion did happen as described because it could not have been fabricated, since its significance could not have been appreciated at the time the Gospels were written. It is significant to us, because it was not significant to the authors of the Gospels. The time of death revealed the truth.

FACTORS AFFECTING DECOMPOSITION

What other factors affect the decomposition of a corpse in a way that can help with time of death determination? We know that humidity speeds decomposition and dry conditions retard it. In buried bodies, it is said that soil pH (the degree of acidity or alkalinity) does not affect the rate of decomposition, an assertion I find hard to believe. It may be true to some extent as far as whole bodies are concerned, but it is not true in relation to skeletons, as we shall see. Some substances, such as quicklime added to the soil, certainly do affect the rate and manner of decomposition.

The time of death of buried bodies can be estimated using entomological (i.e. insect biology) techniques, very much as can exposed bodies, although the fauna involved is quite different. The fauna in the soil beneath a buried body (i.e. a body buried without a coffin) will change in certain predictable ways. First, the existing fauna will decrease, both in number of species and number of individuals. This happens because the decaying body releases decomposition products with such evocative names as cadaverine and putrescine, which are highly toxic. The fauna slowly disappears, a process that takes something of the order of two months. After that a new fauna begins to appear; it will be different from the original fauna and will gradually develop to its maximum extent. The exact order of events will depend on the nature of the soil and the time of year; and it is not possible to be very precise in all cases.

One of the main biological events occurring in soil is the growth of plant roots. It sometimes happens that the roots of a tree penetrate through the skeleton of a buried body and knowledge of the speed of their development will give an indication of time of death. Also, the number of growth rings in roots that had been damaged while the grave was being dug will reveal the time of burial. These methods were used in a case in Bedfordshire in 1978 and it was concluded that the body had been buried three years earlier. Also, the absence of fly pupal cases indicated that the body must have been buried very soon after death.

Taphonomy, the pattern of bone scattering discussed in the last chapter, can also give a general idea of time of death, since a correlation has been found between time of death and extent of disarticulation and scattering of the bones. Bone scattering is said to begin at about five weeks after death, although I do not doubt that much will depend on the time of year in which the person died, as well as on the local conditions, such as the species of wild animal present. Such things cannot be timed to perfection – too many factors and unknowns are involved.

Teeth often acquire a pinkish pigmentation after death, this colour appearing a few weeks post mortem. The discoloration is caused by the accumulation of haem compounds, but how this arises is not understood. It seems that the discoloration usually takes place in victims of violent death. These facts can be useful in time-of-death estimation, but their potential has not been fully developed. Interestingly, the aorta also develops a pinkish colour after death, although this starts days, rather than weeks, after death.

It might seem surprising that all the techniques so far examined are somewhat vague and uncertain, lacking the "Yes" or "No" character that the public in general – and lawyers in particular – expect of science. Are there, then, no such techniques that can be applied to this vexed question of time of death?

Well, yes there are, but they, too, are limited in their own ways. Most can be applied only under special circumstances and, even then, their precision of measurement does not necessarily mean that they will give us a precise answer. If this sounds self-contradictory, let us look at some of these techniques and see what they can do.

Human remains are often found as skeletons. In such cases it is possible to analyse the chemical composition of the bones with a view to establishing time of death. The rate at which nitrogen is lost from buried bones has been used as a time marker in the technique known, wonderfully, as FUN analysis. The letters stand for fluorine, uranium and nitrogen. Nitrogen loss is essentially protein loss, since proteins are the only nitrogenous compounds found in bone. More specifically, most bone protein is collagen, which accounts for ninety per cent of skeletal protein and for about a third of the protein in the body as a whole. It is the main component of tough tissues like tendons and is the substance that yields gelatine when boiled. Its rate of loss, together with the rate of uptake into the bones of the chemical elements fluorine and uranium from water in the soil can give an approximate time since death. It was FUN analysis that enabled Dr Kenneth Oakley and his colleagues at the Natural History Museum in London to expose the Piltdown hoax, demonstrating that the bones were modern and not ancient, as had been claimed.

The nitrogen content of bone is measured using the micro-Kjeldahl technique, in which the dried bone is treated with sulphuric acid to yield ammonium sulphate. This salt is in turn treated with sodium hydroxide to release the ammonia from the sulphate. The ammonia, which is a nitrogenous compound with the formula NH_3, is then absorbed into hydrochloric acid. The residual acid is then titrated. This allows the amount of ammonia produced to be calculated and hence the amount of nitrogen in the bone.

This very precise method of estimating the amount of nitrogen in a bone sample is not matched by a great precision of dating the remains. The reason is that we do not know how much nitrogen is lost from bones over a particular period of time. This is because the rate of nitrogen loss is greatly influenced by temperature and the amount of water in the soil during the period in which the bones had been buried. Nitrogen forms about four per cent of bone by weight and the rate of its loss is also affected by the size of the bone fragments, the smaller fragments losing nitrogen more rapidly. Protein in bones survives for longer in cold, dry conditions, especially if there is little oxygen present and the soil pH is high (i.e. non-acidic). Conversely, it lasts for a much shorter time in wet, warm conditions that encourage microbial growth. Bone will be destroyed very quickly in very acidic soils.

Protein is made up of chemical building blocks called amino acids. These are lost selectively over time. In other words, some amino acids disappear before others. These facts have been used as a technique of time-of-death estimation by archaeologists and attempts have been made to apply them to forensic problems. The technique involves the breaking down of collagen in the laboratory, then identifying the various amino acids found. The basic idea here is that, the longer the post mortem period, the fewer amino acids will be found. Fresh bone yields between ten and fifteen different amino acids, whereas bones that are over a hundred years old have less than seven. The absence of two amino acids, proline and hydroxyproline, indicates that the bones had been in the ground for hundreds, even thousands, of years.

The relative proportions of amino acids in the bone has also been used as a dating technique. Interestingly, a 2,000-year-old sample of bone from Egypt showed that the amino acids present were very similar in relative proportions to a modern bone, with the exception of one acid, threonine. The concentration of amino acids was more than ninety per cent of that in modern, fresh bone. On the other hand, a 5,000-year-old bone from California was found to have lost all its proline and hydroxyproline, and that aspartic acid was present in disproportionately greater amounts. The total concentration was less than one per cent. In general, it was found that the relative proportions of amino acids in old bones was roughly similar to modern bones, as long as more than ten per cent of the original amount remained. At lower proportions the amino acid "profile" will look very different. The reason some amino acids are lost faster than others lies in the fact that they are more easily dissolved in water than others. This applies to proline and hydroxyproline, which, as we have seen, are lost relatively quickly. Some amino acids are more susceptible to destruction by acids than others; this is the case with tryptophan and threonine, for example.

There are three problems associated with the application of this technique to forensic science. First, the obvious reason that it is highly unreliable. Secondly, there are unresolved chemical problems with the method of extracting the amino acids in the laboratory. Thirdly, the timescales involved are too great, rendering it of little use in modern case-work. However, it is not entirely without interest to the forensic scientist, since it can, at least in many cases, distinguish between a recent death and one that happened several hundred years ago. This is important, since the discovery of, say, a 500-year-old skeleton is of no legal interest and need not be investigated by the police. Also, we must not forget that some cases of forensic interest are historical ones, as we shall see more clearly in later chapters.

When bone is exposed under ultra-violet light it fluoresces a blue colour. Older bones have been found to fluoresce a yellowish-green colour; this latter is attributed to the presence of the metabolic products of bacteria and moulds. Other old bones, such as those excavated from an Etruscan site in Italy, fluoresced only very weakly. More recent samples from Mediaeval times fluoresced bluish-white, but more weakly than modern samples and the distribution of fluorescence was patchy;

moreover, it was restricted largely to the centre of the bone. It seems that the ability to fluoresce is lost with the lapse of time from the outside of the bone inwards. It is generally thought that bones that fluoresce across the whole of their cross-section are less than a hundred years old. Older bones, up to 800 years old, will have completely lost their ability to fluoresce.

These findings are very promising, and great efforts are being made to understand the exact relationship between loss of fluorescence and passage of time, since if this could be quantified it might prove a useful forensic tool. However, it is not yet established what it is that fluoresces in bone; some researchers believe it to be the organic (protein) parts, while others believe it to be the inorganic (mineral) part.

This uncertainty introduces another problem. Bones lying in the ground are exposed to the process of remineralization. Such changes in the mineral content of the bone would then confuse the picture, since one would not know whether the presence of strong fluorescence is due to the original mineral components, or to later additions. Furthermore, some of the mineralogical changes that take place in buried bones are well understood, such as the increase in size of the hydroxyapatite crystals, but other changes are not at all understood.

Aminobenzidine is a chemical that stains haemoglobin, the purple pigment in red blood cells. (Blood appears red only when it is oxygenated.) Applying this stain to old bones shows that, generally speaking, bones older than fifty years do not react positively. Nevertheless, exceptions are known, since some bones as old as a hundred and fifty years have exhibited a positive reaction. Confusingly, blood remains have been detected in bones that are as much as ninety thousand years old. Luminol is an organic chemical that reacts with the iron in haemoglobin, and then gives off a striking blue chemoluminescence when mixed with an oxidizing agent. It is widely used for detecting trace amounts of blood, as we shall see later in this book. More recently, the luminol test has been suggested as a possible tool for estimating the post mortem interval of skeletal remains, with the aim of, at the very least, identifing those remains that are of medico-legal interest, i.e. less than one hundred years old. Chemoluminescence visible to the naked eye has been observed in bones that were up to fifteen years old,with a steady decrease in luminescence with increasing post mortem interval. Bones that are more than eighty years old no longer show this bioluminescent effect. However, this technique, too, is severely limited in its applicability.

There are other dating techniques, many of which are at present in greater or lesser degrees of development. One most promising line of investigation is the analysis of the presence of the body's decomposition products in the soil. It would be tedious to catalogue all the work being done in this area, but suffice it say that many, varied lines of research are being pursued, and one example will make the point. The increasing concentration of inorganic ions (i.e. ions lacking carbon) in the soil has been developed in the United States. Interestingly, a way round the common problem that so many natural processes are dependent on temperature was found. It was calculated that the concentration of inorganic ions* increased, not simply in relation to time, but in relation to accumulated day degrees (ADDs), which is the product of the temperature and the number of days. (This method is used in other techniques, such as the study of maggot development above.) In other words, the same change would take place over ten days at 20°C, as would take place over twenty days at 10°C, the ADD being two hundred in both cases. So, some kind of indication is given, but not in absolute time terms. Peaks of decomposition products could be shown to occur at predictable ADD intervals. Moreover, the time scales involved, whichever way one "read" the ADD, were applicable forensically, being of the order of hundreds of days.

These results, although interesting, cannot be applied universally, since the nature of the soil and, hence, the rate of loss of decomposition products, will vary from place to place. This is a problem that affects all techniques involving the soil surrounding the body.

Having come this far, it may seem that there is no absolutely reliable time-of-death dating technique. One problem that bedevils all the methods discussed above is the fact that they are all temperature-dependent. Before we can interpret our findings, we have to reach some conclusion about the temperatures prevailing during the relevant period. This is no easy matter.

So, what next? People are accustomed to hearing about such success stories as the use of radiocarbon dating of the Turin Shroud, which was shown to have been of Mediaeval, not ancient, origin**, that I am often faced with a look of disbelief when I say that dating the time of death is so difficult. These days everyone is familiar, not only with radiocarbon dating, but with other such high-tech methods as accelerated mass spectroscopy, thermoluminescence and electron spin resonance, all of which sound wonderfully infallible as dating techniques. Why, then, are there no fail-safe methods of determining the time of death? Is it true that there are no available techniques that do not depend on phenomena of natural change that take place at a known rate and totally independently of temperature? The answer is, no, it is not wholly untrue, but such techniques as are available have problems of their own, too.

First, a technique like radiocarbon dating comes into its own only when we are concerned with very long periods of time, because it cannot be used to detect or interpret the changes that occur over short, more recent periods of time. It would be like trying to time a race that lasts for a few seconds using a clock with only an hour hand, or like weighing a button using a steelyard. Even over long periods of time, radiocarbon dating has correspondingly large margins of error, although, over the time periods in question, they may not appear important to us. For these reasons it cannot be used as a forensic dating technique. It is generally agreed that radiocarbon cannot be used to date events that took place more recently than four hundred years ago.

Radiocarbon dating exploits the fact that carbon-fourteen (14C) decays (i.e. becomes non-radioactive) at a known rate. It has a half-life of 5,730 years, the half-life being the time it takes for half the nuclei to decay. It is this change that can be used as a measure of time.

Yet there are other radioactive techniques that can be used in forensic science, two of the most important being the consequences of two of the great disasters of the twentieth century: the Second World War and the explosion at the Chernobyl Nuclear Power Station in the Ukraine. Both events produced peaks of radioactivity, the decline from which can be used as date markers since those events took place.

The numerous nuclear arms tests that took place after the Second World War increased the levels of radiocarbon, or carbon-fourteen (14C), in the atmosphere. The peak was reached in 1963 and the levels have been declining ever since. This radioactive carbon has ramified throughout the living world and its presence and exact level in human bones can be used to determine whether such a person died after 1950, the year in which nuclear testing began.

Levels of other radioactive isotopes, such as tritium (3H – a radioactive isotope of hydrogen), caesium-137 (137Cs) and strontium-90 (90Sr) have also increased since the Second World War. These have much shorter half-lives than carbon-fourteen, it being 12.4 years in the case of 3H, for example. After the Chernobyl accident, further amounts of radioactive isotopes were released and are of potential use in forensic dating. Interestingly, studies have shown that 90Sr levels were high in samples taken during recent post-mortems. Comparable samples from mediaeval times showed lower levels, but it was somewhat surprising to find any 90Sr in them at all. This was interpreted as being contamination from groundwater, generating another complicating factor.

Let us look at one last, relatively recent, development. It was found that the rate that potassium enters the vitreous humor (the jelly-like substance filling the eyeball behind the lens) takes place at a constant rate and, moreover, is a change that is independent of temperature. The potassium derives from the breakdown of the red blood cells. Have we, at last, found the ideal indicator of time of death? Alas, no. First, like the conventional post mortem changes, it is detectable only during the first few days after death. Moreover, I am far from convinced that the process is as wholly independent of temperature as has been believed, but I keep an open mind about it. Time will tell.

However, since estimating postmortem interval (PMI) is one of the keystones of a forensic investigation, it will come as no surprise that PMI estimation is proving to be a fruitful area of research for scientists in many varied disciplines. The advent of novel technologies has opened up new avenues of research. Some potential lines of enquiry involve the use of electric impedance spectroscopy, a well-established method for characterizing the electrical properties of materials. As postmortem interval increases, tissues in the body gradually undergo autolysis and decomposition, which would be expected to increase the permeability of cell membranes and decrease the capacitive reactance of those tissues leading to a decrease in electrical impedance. Studies are ongoing to investigate how such impedance changes might be linked to PMI. Other lines of research for estimating PMI include the use of the cellular content of cerebrospinal fluid, quantifying the amount of melatonin in a corpse, photometric measurements of the colour changes that take place during livor mortis as a function of both pressure and time, immunohistochemical techniques evaluating the changes in composition of specific cells with time since death, and using Fourier transform infrared spectroscopy to measure changes in specific tissues. At present, none of these methods are widely accepted but several show promise.

I hope this chapter has done more than describe a set of techniques. In addition to that, my aim was to show that science is far from being the idealized, almost magical, process many non-scientists seem to think it is. We are ignorant of many things. Simple questions do not necessarily have simple answers. The cure for some cancers has been found, whereas the cure for the common cold still eludes us.

The faith that many lay people have in science and scientists, while flattering, is often unjustified. Scientists, not least forensic scientists, are faced with many complex problems that often defy easy solution. Some members of the public may find this surprising, while at the same time find nothing strange in the fact that scientists are continually carrying out research. If scientists had all the answers, they would not be so busy in the laboratory and in the field.

It is, perhaps, true to say that forensic scientists are the least "academic" of all scientists, by which I mean that they are much closer to the real world and that their conclusions are subjected to much fiercer public scrutiny. Their findings are exposed to immediate testing in an uncompromising environment. All the greater, then, is the pressure upon them to produce answers to questions, solutions to problems. And so science advances: by not knowing the answer to begin with, by thinking how to cope with a problem in ways that have not been thought of before; by asking questions that normally issue from the mouths of babes.

Solutions, sometimes, can be remarkably easy. When the notorious 1930s murderer Dr Buck Ruxton killed his wife and maid, cut up their remains and dumped them in a Scottish ravine, various methods were used to determine the time of death of the two women. In the end, the question was answered very simply: the date on the newspapers in which Ruxton wrapped the body parts told the police all they needed to know.

---

*An ion is one of the two electrically charged "halves" of an inorganic compound. The word "inorganic" refers to a compound lacking in carbon, with the exception of such common compounds as carbon dioxide and various carbonates.

** The date of the Turin Shroud has not, in fact, been conclusively demonstrated. (See Chapter 8: *Words and Images*.)

CHAPTER FOUR

A QUESTION OF IDENTITY

Art thou anything?
Art thou some god, some angel, or some devil,
That mak'st my blood cold, and my hair to stare?
Speak to me what thou art.
William Shakespeare
Julius Caesar *Act IV, Scene 3.*

The young CID officer from Yorkshire, smart and dapper, was describing the scene to me. The body of a murdered woman had been found in a wood, but she was still unidentified.

"If only we knew who she was, doctor, we'd be able to make progress," he said.

He sat, frowning, deep in thought, no doubt thinking of ways to answer that question. It was to be a long time before he had his answer. Eventually, after a long police investigation, the identity of the woman became known. She had been a prostitute, killed by one of her customers. Her departure from this mortal coil caused no-one any concern, not even those who knew her well, her many devotees declining to come forward for fear of damage to their own reputations. Such, alas, is human nature; self-interest so often overrides concern for others, even to the extent of allowing a person to disappear from the face of the Earth, without any indication that they have gone.

We have dealt with the "where" and the "when" in earlier chapters; now we ask "Who?". Who was the victim; and, equally importantly, who was the criminal – the classic question, "whodunit?". In the case above, the dead woman was identified as a result of tireless and protracted police inquiries, but forensic assistance for identification is frequently needed by the police and, happily, many techniques are available.

We started this book by looking at Locard's Principle, the basic tenet of forensic science; but there are other tenets of great importance. One of these is the Principle of Individuality, which states that no two objects are identical. No two people, no two documents, no two skulls, no two fingerprints, even those from the same person, are *identical*. This does not mean that they are, perforce, distinguishable in practice, but they are distinguishable in principle. Forensic techniques are constantly being developed to increase our powers to distinguish between things or people.

IDENTIFICATION

The first attempt to identify people on a rational, scientific basis was developed during the second half of the nineteenth century by the French forensic scientist, Alphonse Bertillon. Anthropometry was a system based on a number of measurements of key facial and body features. It was a good beginning and it had some successes, as when Bertillon asserted that a criminal named Dupont was the same person as a man named Martin – Dupont later admitted that he was, indeed, Martin.

In spite of its successes, anthropometry suffered from a number of flaws, the most important of which was the fact that it was impossible to assess the probability of two people having the same sets of measurements. Also, it could only be used comparatively, to say whether a person had similar features to a known criminal. For these reasons, as well as the cost involved in training police officers to apply it routinely, anthropometry gradually fell out of favour.

Bertillon developed another technique, called *portrait parlé* – a "word" or "speaking" portrait. This system involved the use of various forms of the facial features and building them up to make a "picture" in words. For example, the forehead might be broad or narrow, the nose high-bridged or concave and so on. The various states in which each facial feature could occur were much more numerous and fine-tuned than the example given in the last sentence, but it serves to make the point. It was useful in enabling police officers to describe a person in a way that would allow others to form a very good idea of their appearance.

The obvious step from a "word" picture to a real picture happened when the Identikit was invented by Hugh McDonald in California during the Second World War. This system employed a number of transparencies of different forms of mouth, nose, chin, etc., which could be superimposed on one another to produce a facial representation. The number of variations of each feature differed from feature to feature; for example, there were one hundred and two eye forms, but only fifty-two chin forms. Nevertheless, Identikit proved to be a valuable method in producing a more or less accurate visual representation of a face from a description given by a witness. Publication of the Identikit in newspapers has resulted in the identification and arrest of many a criminal.

The logical next step came when Jacques Penry in Britain developed the technique known as Photo-FIT – the last three letters standing for Facial Identification Technique. This was essentially the same as Identikit, but used photographs rather than drawings. It was also more detailed, including, for example, one-hundred-and-one mouth forms and eighty-nine nose forms, compared with Identikit's thirty-three and thirty-two, respectively. It was possible to construct fifteen billion different faces using Photo-FIT.

With the advent of computers came the development of E-FIT (Electronic Facial Identification Technique) and it is now possible to reconstruct faces in three dimensions and to introduce other kinds of feature, such as colour and clothing. Computers have also played an important role in the development of techniques of identifying, not only the living, but the dead, as we shall see later in this chapter.

MATCHING PRINTS

Fingerprint analysis is the classic tool of crime detection, yet it is a technique that is little understood, not only by the public at large, but by the law courts as well. Surprisingly, fingerprint specialists themselves often give the impression that they are not wholly aware of the full power of their methods.

Fingerprints are made up of many ridges, arranged in rough, concentric circles. Some of the ridges end abruptly (ridge endings), others end in forks (bifurcations) and yet others fork and close up again (enclosures). Collectively these ridge characteristics are known as minutiae. The full number of permutations is infinite, hence the perfectly justified belief that no two individuals have identical fingerprints. The various features of a fingerprint, together with their relative positions on the print, are what enable the fingerprint examiner to identify a particular set as belonging to a specific individual.

In Britain, the traditional way of doing this is to match sixteen points of the print with the suspect's fingerprint; if they match exactly, then an identification can be made. Yet there is no reason to suppose that the number sixteen has any special significance. The number was not arrived at through any process of logical reasoning, it simply appeared to be a safe number of points upon which to arrive at a conclusion. In Paris, seventeen points of congruence are required to prove identification, whereas only twelve are required in the rest of France, as advocated by Locard himself. Again, only twelve points are required in Australia and New Zealand, whereas only eight points of agreement are used to prove identity in Turkey. In India the figure varies from six to twelve, depending on the particular state. The United States no longer has a fixed number of points.*

Does this mean that standards are highest in Paris and Britain and lower elsewhere? No; in fact, in one sense, it may mean the exact opposite. To explain this apparently self-contradictory statement it is necessary first to look at the way fingerprint analysts actually identify fingerprints.

The currently accepted technique is known as ACE-V (Analysis, Comparison, Evaluation and Verification). With the advent of computers, many countries have adopted the practice of scanning and digitally recording sets of fingerprints into vast databases** that can be checked for matches using an automated search facility. The computer uses a search algorithm, based on criteria that can be adjusted by the operator, which calculates the degree of correlation (usually reported as a "score") between the location of ridge characteristics and their relationship to other minutiae for both search and file prints. Thousands of fingerprint comparisons can be made in one second. When the search is complete, the computer will produce a list of file prints, or "hits", that have the closest correlation to the search print. However, it is important to note that, in all cases, a trained human fingerprint examiner has the final say as to whether a print is a match or not.

The human examiner usually begins by studying the prints as a whole. He then makes comparisons between equivalent parts of the prints to see whether they match. Next, having made up his mind that the two prints came from the same individual, he draws up a list of the matching sixteen points and presents his report. Note that I have written that the identification is made first, the listing of the points of congruence being made after the fingerprint analyst has satisfied himself of his identification. I do not mean by this that these examiners are in any way dishonest about their conclusions. What I do mean is that they actually arrive at their conclusions on the basis of their experience and knowledge and not on the basis of the counting of points of similarity. The following results from a Home Office study should make the point clear.

In that inquiry, pairs of prints from the same individual were sent to fingerprint examiners throughout England and Wales. In addition to the true pairs of prints, one pair of unidentical prints (i.e. not from the same individual), which were altered to make them appear to be the same, were sent to the examiners, who were asked to determine whether the prints in each pair came from the same individual and to give the number of points of similarity.

All the examiners saw through the deception of the doctored pair of prints, none misidentified the closest pair of similar prints that could be produced by any fingerprint bureau in the country. On the other hand, they disagreed widely among themselves as to the number of points of similarity between the genuine pairs; in one case the number of perceived points of resemblance varied from eleven to forty. A similar study carried out in the USA was reported in 2011. There, 169 latent print examiners each compared roughly one hundred pairs of fingerprints from an overall pool of 744 pairs chosen to include a range of quality of print, as would be found in normal casework. Only five examiners made a single false positive error (that is declaring a match where none existed), giving a false positive rate of 0.1%, compared to an overall false negative rate (declaring no match between prints known to have come from the same individual) of 7.5%, but as in the Home Office study, examiners differed widely on whether a particular print was suitable for reaching a conclusion or not.

The conclusion that can be drawn from this study is very clear. Fingerprint examiners are extremely good at matching two prints successfully, but their ability has very little to do with the counting of points. Although this may sound a rather startling conclusion, in fact, it should not surprise us, for recognition and description are quite different things. You and I may be able to recognize our friends and relations with ease, but, if asked how we were so sure that they were not other, very similar people, we would probably falter and hesitate. We are sure of our identification and that's that. We know, but we do not know how we know. The analytical ability to "dissect" a pattern and produce a description is a power that not many people possess.

So, what is the answer to our question, "Are standards higher in countries that demand a larger number of matching points?". The answer must be that, while a sledgehammer will crack any nut, a nutcracker would do just as well. In other words, the desire for a high number of matching points does not raise standards, it merely makes it look as though one is erring on the side of safety. Since we know that fingerprint examiners identify fingerprints on the basis of their experience and not on the basis of point-counting, this excessive caution is not only unjustified, it excludes good evidence from the courtroom. This is because a fingerprint examiner may be sure that two prints came from the same individual, but may not be able to find the required sixteen points of similarity. It must be remembered that many fingerprints found at scenes of crimes are only partial prints.

It is odd that fingerprint examiners themselves have not been able to accept these conclusions, perhaps fearing that their subject would be seen as being unscientific if they did so**. But this is to miss the point completely. Although one cannot attach any great significance to the number of matching points in two fingerprints, one can measure the probability of a fingerprint examiner arriving at the correct answer. This is perfectly respectable and scientific. It does not matter how fingerprint examiners arrive at their conclusions, as long as studies show that they are almost always correct.

It has been pointed out many times, notably by Dr Ian Evett and his colleagues, that one should concentrate on professional standards, that is, on the demonstrated ability of fingerprint examiners, rather than on the counting of matching points. This is sound advice and I hope that both fingerprint examiners and the courts take it on board.

LOCATING AND LIFTING PRINTS

Before a fingerprint can be used as an identification tool it has to be located and preserved. Visible prints, as their name suggests, are clearly there for all to see. These will have been made when the fingers touched a surface after coming into contact with a coloured material such as blood, paint or ink; or they may have been left as impressions on a soft material such as soap, wax, or even dust, this latter group being termed plastic prints. However, the majority of prints left by an individual will be latent prints, invisible to the naked eye. The examination of fingerprints is usually done at the scene and, until recently, British police forces did not "lift", i.e. remove an impression from the scene, in case it damaged the evidence. Despite the fact that other countries routinely lifted prints from the scene to the laboratory, this attitude continued in Britain until 1970, when "lifting" became acceptable.

There are various ways in which a print can be enhanced or lifted. Anyone familiar with cinematic whodunits will be familiar with the use of carbon or aluminium powders to to "dust for prints" in order to enhance the impressions for subsequent photography or lifting. According to the nature of the surface upon which the print has been left, different kinds of dye might be used to render the impression clear enough for analysis. Nowadays, investigators routinely use hand held Alternate Light Source

(ALS) devices, such as the RUVIS (Reflected Ultra-violet Imaging System), in combination with a variety of chemical sprays that induce fluorescence for detecting latent fingerprints.*** The enhanced print should be carefully photographed, both in close-up and in an overall view to show the print's location to other evidence at the scene, before any attempt is made to "lift" the print.

"Lifting" itself is a very straightforward operation, although it has to be carried out carefully. It is done by placing low-adhesive tape on top of the enhanced print, then carefully removing it and sticking it down again on a piece of white card, on which the details of the case are recorded. Commercial gel-lifters of varying sizes are also becoming popular.

Fingerprints lifted from a crime scene are usually not in the most perfect condition, but if digital photographs are taken, or the print is scanned into a computer file, digital imaging software can be used to improve the image, which can then be fed into an automated fingerprint identification system.

Taking fingerprints from known individuals – living or dead – is again a reasonably straightforward matter; everyone is more or less familiar with the traditional process of coating the fingers with ink and taking an impression. A certain protocol being followed when doing this; for example, there was an established order in which the prints from the different fingers were taken and deposited on a form with a number of boxes, each box being reserved for a different finger. Occasionally difficulties arose, especially when fingerprinting the dead, since the fingerprints may have been damaged as a result of mummification, decomposition or excessive wetting. Special techniques are available to deal with such situations, although it is usually a matter of intelligent improvisation. Today, as anyone who has recently travelled to the United States can attest, fingerprints are routinely taken using a scanner, one simply presses the thumb and fingers on to a glass plate and a digital scan is made and added to the fingerprint database.

The huge number of cases in which fingerprints were used to identify a criminal makes it difficult to choose an example as an illustration. In view of this, I will take as an example the very first use of fingerprints in a criminal case, which took place in Argentina in 1892.

Two children, who lived with their mother in a shack near Buenos Aires, were found bludgeoned to death. The mother was slightly injured, but not badly hurt. The police suspected a man named Velasquez, since the mother Francesca Rojas had accused him of the crime. However, fingerprints left on blood splashed on the shack door matched those of Francesca herself. At that time, fingerprints would probably not have been accepted as evidence in court, but, when confronted with the evidence, Francesca Rojas broke down and admitted that she had murdered her children. The reason for her foul deed was that her lover said he would have married her, if only she had had no children.

It is not only the tips of the fingers that can leave tell-tale signs; the whole hand and the bare foot can do so too. Various systems have been developed to identify individuals using hand and foot prints.

SNIFFING OUT EVIDENCE

Everyone knows that dogs have an uncanny ability, through their sense of smell, to recognize individual human beings. Their ability to follow a scent over great distances, over wet fields in the pouring rain, is a matter of common human knowledge. They can also transmit this information to us; their subsequent behaviour leaves us in no doubt that they have successfully accomplished the task they had set out to do.

It may come as a surprise to learn that in Britain, a great dog-loving nation, dogs are not used to identify people in criminal investigations. The evidence of a dog is inadmissible as evidence in British courts, although one or two exceptions to this general statement are known. It is true that dogs are used to find people and dead bodies, but this is not the same thing as identifying a person. When the police search for a missing person they use dogs to track them down. When the person is found that is the end of the matter. The dog's discovery of the person is not used as evidence in any way; the animal was simply being used as a tool to find the person. Should the identity of the found individual be in doubt, then other ways of establishing his or her identity are used.

Similarly, dogs are often used to find people lost on mountains or buried beneath the rubble of a landslide or an avalanche, but, again, the dog is merely a device to achieve an end, much as a police officer might use a car to go from one place to another.

If, on the other hand, a dog snarls at an individual who is suspected of having burgled the premises of the dog's owner, this is not seen to be evidence of the person's identity; in other words, it cannot be used as evidence that the man was the burglar. This is the case even if the dog barked at the man after being allowed to sniff, say, a glove that was left behind by the burglar.

Why is such evidence not allowed in British courts? It is acceptable evidence in Holland, Germany, Hungary and some parts of the United States, but not in Britain. Could it be that the ability of dogs to identify people is exaggerated, that they are not really as proficient at this task as they are claimed to be?

Let us look first at the evidence for the power of dogs to recognize people through their sense of smell, before answering the question about the legal use of their abilities. The police force that has done more than any other to develop the use of dogs as identifiers of human beings is the Rotterdam police in Holland, where the use of dogs is now routine. Experiments carried out at Rotterdam have contributed immensely to our knowledge of the smelling abilities of dogs. It is worth looking at these experiments and their results.

To test the ability of a dog to identify a specific person, a cloth that had been previously handled by that person is given to the dog to smell. Together with a number of other people, the man who had handled the cloth is required to have a shower, using a particular brand of soap. They are then required to wear identical garments that had been washed in the same way and with the same washing powder. Next, they each stand in a separate cubicle in a row inside a room. A screen in front of each man prevents the dog from seeing them. Fans, one behind each man, are switched on.

The dog is now taken into the room, together with its handler. The handler is not told of the position of the man who had handled the cloth, so that he cannot influence the dog's choice, however unconsciously. The dog, having smelt the cloth, is walked back and forth in front of the row of cubicles by the handler. Eventually, usually very quickly, the dog stops in front of one cubicle and barks. It has identified the person and hardly ever makes a mistake.

The dog is taken out of the room and the people in the cubicles change positions. The man playing the part of the culprit is now in another cubicle and the experiment is repeated. Again, the dog finds the right man. Next, another experiment is conducted. Cloths are handed to each of the people involved in the experiment; they handle them and then place them in special jars, with each cloth in a separate jar. The jars are placed in a row in the experiment room and the dog and handler come in. The dog sniffs each jar in turn and then identifies correctly the jar with the right cloth. As with the earlier experiment, the jars are moved around in the absence of the dog, who returns with his handler and correctly identifies the cloth.

These results are very impressive, but, to my mind, the results of the next experiment are the most impressive of all. The jar with the "right" cloth is removed completely, leaving all the other jars, plus another to keep the number constant. What will the dog do now?

As with the other experiments, the dog is led by its handler into the room. The dog sniffs each jar in turn. It is puzzled. It starts again, sniffing each jar diligently. It stops and looks up at its handler and then looks back at the jars. It then starts to whine to its owner and walks away from the jar; no doubt it feels that it has failed in its task.

But it has not; it has succeeded brilliantly, for the dog has not chosen a second best, a nearest odour to the one it was seeking. The smell was either present in one of the jars or it was not. It is as simple as that. The dog would not identify a false jar even to please its handler; it would rather fail than do that.

A great deal of experimental work of this kind has been carried out, but the above experiments suffice to show that the sniffing abilities of dogs and their power to identify individuals with almost unerring accuracy are very great. In Britain, my colleague Dr Barbara Sommerville (now retired) of the Department of Clinical Veterinary Medicine at Cambridge University, conducted similar experiments and was able to demonstrate that dogs can even distinguish between identical twins. In Norway, scientific investigations showed that dogs that have been oriented at an angle of 90° to the direction in which a person has walked will turn in the right direction within two to five seconds. Some dogs have been able to track people forty-eight hours after they had crossed a field in winter.

All this shows that there can be no doubt that dogs have formidable powers to find and identify people. Such powers ought to be harnessed for the benefit of the criminal justice system in Britain, but the courts continue to resist their use.

Why? The reason is that, if one accepts the evidence of a dog it would be tantamount to saying that a dog is the equal of a human being. One judge is reported to have said that he would not accept the evidence of a dog against the evidence of a man and many commentators have ridiculed the idea of a dog appearing in court answering the questions of a magistrate. It is seen as a fairy tale or science fiction. "What next?", such people have asked, "Are we to have cats and horses and cows giving evidence in courts of law? Such ideas are ridiculous and not worthy of further discussion."

And so they are, but such things are not being proposed. Nobody is suggesting that dogs should give evidence in court, only people. Dog evidence is no different from any other kind of evidence; courts have accepted the evidence of blood, glass, textile fibres and a whole host of other things, but nobody, to my knowledge, has ever said that they would not accept the evidence of a fibre or a piece of glass in favour of that of a human being. Nobody suggests that the evidence obtained through a chemical reaction in a test tube requires us to accept the evidence of the test tube or that the test tube should stand in the witness box and give evidence. The evidence lies in the interpretation by a human being of the analysis of the blood or glass or fibre or whatever it happens to be. It is the chemist or the biologist or the fibre specialist who gives evidence. And so it is with dog evidence. It is the person knowledgeable about dogs and dog behaviour who analyses and interprets the dog's behaviour, on the basis of scientifically acquired knowledge, and who stands in the witness box and gives evidence.

It is a simple case of being at cross purposes. Dogs are much more like human beings than are test tubes or pieces of glass. When the idea of dog evidence is raised, some people, who have not thought the matter through, think it amusing and unreal that a dog should give evidence. This confusion would, indeed, be amusing were it not for the fact that it is depriving the legal system of a very powerful line of evidence and the police of a very effective weapon in their crime-detection armoury.

I believe that the use of dog evidence in British courts would be a great step forward in the fight against crime. The value of dog evidence, I make no apology for repeating, would be enormous, especially in cases of burglary. Burglars almost always leave behind a trace of their body odour, which could be identified by a dog and used as a tool against such criminals. The intellectual confusion surrounding the use of dogs as evidence must be seen as a fallacious and simplistic argument against a highly desirable new resource in crime detection in Britain.

Attempts have been made to produce a machine – an electronic nose – that can do what a dog does. These devices have been very successful in determining whether a food product, such as wine or cheese, is fresh and in fit condition to be consumed. In terms of their application to criminal investigation, most work has concentrated on developing bomb and drug 'sniffers'. For person identification from scent a dog is still the more reliable tool.

Edmond Locard waiting to testify at a criminal trial.

Sir Arthur Conan Doyle: a pioneer of forensic science.

O.J. Simpson.

"Saint Martin and the Beggar" by El Greco. Note the
unnaturally attenuated form of the beggar.

Forensic scientists collecting evidence at a murder scene.

A mass grave being exhumed in Sarajevo, Bosnia, 1998.

An aerial view of a murder scene.

Electron micrograph of two-hour old greenbottle maggot.

"Christ on the Cross" by El Greco.

Mr Alvan T. Marston demonstrates that the Piltdown cranium is human and comparatively recent in origin.

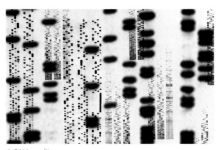

A DNA profile.

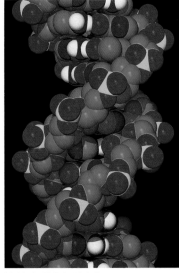

Model of a DNA double helix.

Medallion of Philip II
of Macedon.

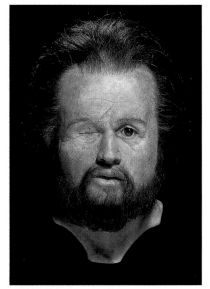

Richard Neave's reconstruction of
King Philip's face.

The Romanov family and attendants.

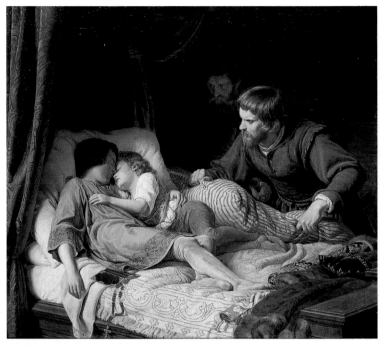

"The Princes in the Tower" by Hildebrandt.

LEFT: Reconstruction of Karen Price's face. RIGHT: A photograph of Karen in real life.

A Bloodhound: dogs have an unerrring ability to find and identify humans.

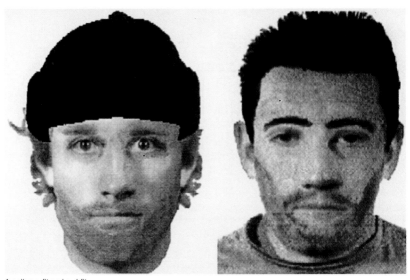

A police e-fit and c-d fit.

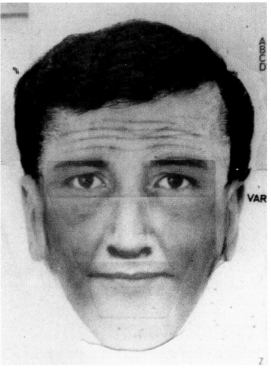

A police photofit.

TRACKING THE CULPRIT

We have seen how a dog can identify a human with unerring accuracy, although the animal does not know how it does it. Moreover, beyond realizing that certain odours are involved, we, too, do not understand fully how the dog does it. This in no way undermines the validity of the efficacy of the dog as a tool. Unfortunately, it is often stated that, if the mechanism underlying a piece of behaviour is not understood, then it cannot be considered "scientific". This is not true, since the proof of the pudding is in the eating, not in how it is made.

There are many techniques waiting to be used forensically, but, because we do not understand how they work, they are largely ignored. One such technique is tracking – not by dogs, but by men.

During the 1920s in Egypt a postman was murdered as he was walking between two villages in the desert. He was shot dead, the bullet passing through his head from right to left, although there was no trace of the bullet itself. The description of the wound in the post-mortem report suggested to Sir Sydney Smith, who was consulted on the case, that the bullet came from a high-velocity weapon, such as a .303 rifle. At first, no motive for the murder could be discerned.

Sir Thomas Russell, Commandant of the Cairo police, had great faith in the tracking skills of the Bedouin. They were brought in and asked to find any tracks that might shed light on what had happened, although the police had already made a search for a track, but had found none.

The Bedouin trackers did find some tracks – of a man wearing sandals, they said. The tracks led to where the body had lain on the sand. The trackers then traced the footprints back to a place about forty yards away; here, they said, someone had knelt down. One of the Bedouin picked up something from the sand and handed it to the police. It was an empty .303 rifle cartridge.

The trackers then said that the man who had left his tracks took off his sandals after inspecting the victim's body, then ran barefooted to the road. They followed the trail, which led to a fort in which were stationed six members of the Camel Corps. There the trail stopped.

Clearly, one of the men in the fort was the murderer, at least according to the Bedouin. The police arranged a march across the desert, in which all six Camel Corps members, as well as a number of other men, took part. The Bedouin examined the tracks and pointed out one set as belonging to the murderer. The experiment was repeated – with the same result. A third time the experiment was carried out and, again, the Bedouin picked out the same man's tracks. A fourth and final test was carried out, this time without the man whose tracks the Bedouin always picked out. After their examination of the marks in the shifting sands, the Bedouin said they could not find the tracks that resembled those near the murdered man; and they insisted that the man could not have taken part in the march.

Although the police were now fairly sure that they had their man, trackers' evidence is not usually accepted in court. Everything now depended on the evidence of the cartridge, which was sent to Sir Sydney Smith. Samples of cartridge cases were obtained from the rifles of all six men – six rounds were fired from each rifle. These were examined under the microscope and compared with the one found at the scene. From various marks on the cartridge cases, each uniquely caused by the rifle from which they were fired, it was possible to say which rifle had been used to murder the postman. It was the rifle belonging to the man whose tracks the Bedouin had picked out.

In the words of Sir Sydney Smith: "Two sciences, one very ancient and the other very modern, led to exactly the same conclusion." Ballistics and Bedouin tracking both identified the same individual.

FORENSIC ANTHROPOLOGY

Physical anthropology – as opposed to social anthropology – is the study of the variation in the anatomy of the human body, especially the form of its bones. Its aims include the determination of the sex, age, stature and general physical condition of bones, particularly those that have been excavated during archaeological or palaeontological work. Forensic anthropology is essentially the same kind of work in a legal context.

When the bones of a totally unknown person are discovered, attempts are first made to establish what kind of person they were. Were they male or female, old or young, white or black? When an initial physical description of the person is made, one can then set to work trying to find a specific identification and to answer the question: "Who was he or she?"

Establishing the sex of the skeleton is usually the most straightforward part of the first stages of the investigation, unless the person in question was a child or, at any rate, a youngster still in the process of growing to adulthood. The fact that sexual differences are not as readily apparent in very young people as they are in adults makes the determination of sex in the very young problematic. In adults, various bones usually indicate the sex of the deceased very clearly. The most obvious of these is the pelvic girdle, which has a different and very characteristic form in adult males and females, related to the child-bearing ability in the female.

However, other bones, such as the skull and the long bones, can also be used to determine the sex of the skeleton. Clearly, the more complete the skeleton, the more accurate one can be; and it is seldom difficult to determine the sex of a complete adult skeleton. The pelvic girdle or the skull alone can be used with a high degree of certainty, but the use of them together increases the confidence that one has made a correct determination.

Conversely, estimating the age at death of a person is easier in youngsters than it is in adults, because there are many more changes in the morphology and anatomy of children. So, such events as when bones fuse or when teeth erupt can be used with a reasonable degree of accuracy to determine age. In adults, such changes do not occur, so age determination is more difficult. The degree of dental wear, the extent of the deposition of cement on the teeth, as well as certain changes in the bones – such as the degeneration of sternal ends of the ribs with age – can be used tentatively to determine age and, more often than not, it is a matter of the experience of the particular anthropologist, whose work in this area may be more of an art than a science; and it is no less worthy for that.

The height of an individual during life is fairly easy to determine when a complete skeleton is available. In such a case, measurement of the bones, plus the addition of a "correction factor" to adjust for the presence of soft tissues, will determine stature. Problems arise when the skeleton is incomplete. Individual bones will give an indication of height, but the reliability of the information differs from bone to bone. For example, the bones of the lower limbs are a better guide to height than the upper limbs.

The skeletons of young people present difficulties in this area, if only because of the fact that different children grow at different rates. In fact, in order to determine the stature of a child from a skeleton, it is necessary to know both the age and the sex beforehand, which is not always possible and is usually difficult, at least as far as sex is concerned, as we have seen.

Other problems arise in the determination of stature. The base-line data collected from living people is often itself incomplete or erroneous. For example, measuring a person who habitually adopts a slumped posture, or an aged person (height is reduced with age) may result in misleading interpretations. Even the time of day on which the measurements were taken may make a difference, since people are shorter in the evening than they are earlier in the day.

Next comes the question of race or ethnic group – the most problematic question of all. The problem here is many-faceted; first, because the concept of race is not clearly defined and differs from person to person. Also, much more intermarriage between different racial groups occurs today than ever before, confusing the question further. Nevertheless, certain generalizations can be made about differences in the morphology of the skull between, say, Europeans and black Africans. These have to be interpreted with care by an anthropologist familiar with at least one of the racial groups concerned.

More general conclusions about a person's appearance or habits can be deduced by other, more obvious, means. Remains of clothes or shoes, the colour of the hair, the presence of such things as jewellery, watches or keys, can all help with identification. The erosion of the teeth may suggest a particular kind of diet; heavily eroded teeth may suggest a diet that included much gritty material. Even the destruction of identifying features may suggest an identity. When Dr Buck Ruxton removed the identifying features – eyes, ears, fingertips, teeth etc. – of his victims the police suspected him all the more, as one who had medical knowledge and who was aware that such features would assist the investigators.

The patterns of the frontal sinuses – the hollows at the front of the skull – are known to vary from person to person; and their description can be an aid to identification. Various diseases present in life may leave their mark on the skeleton; and their presence will add another aspect to the description of the living person. For example, degenerative joint diseases may leave their mark as massive new bone formation. The bones of one who habitually rode horses during life can be identified as such from a study of their skeleton.

Once a general description of a person is made, the question of specific identification can then be tackled. In this context, the teeth are particularly useful, not only because they are the most persistent and indestructible of all body parts, but because they vary so much between individuals. Also, the fact that dental records (i.e. the history of their dental treatment, including fillings) of most people are held by dentists, means that there is a reference point that can be used to check the identification.

Dental identification is the norm in mass disasters. For example, the identification of the victims of the King's Cross Underground fire in London was made by the comparison of dental records with the teeth of the deceased. In such a situation the method is particularly effective, since the records of missing people (i.e. people who might have been travelling on the train) could be consulted and compared with those of the deceased individuals.

In practice, the identification of human skeletons can be wildly successful, or disappointingly inconclusive, according to the nature of the evidence to hand. I was once involved in a case in which the identification of the skeleton was a matter of great importance. Many specialists were called, but the age, sex and race of the victim were never definitely established.

Before we look at further methods of personal identification, let us examine a case of old bones, whose identity is still a matter of dispute. We have already touched upon it; it is the story of the Princes in the Tower.

OLD BONES

When the bones of two children were discovered buried deep at the foot of the stairs in the White Tower, it was immediately concluded that the skeletons were those of the princes. In 1933 the skeletons were subjected to a forensic examination. A basic flaw of this investigation is that it was carried out on the assumption that the remains were those of the two princes, whereas this was the very question that needed to be answered. Nevertheless, the findings were of some interest. It was concluded that the skeletons were those of young people of roughly the ages of the princes when they died, although margins of error were given. The sexes of the skeletons were not established nor was their age (i.e. the period of history in which they died).

Were these skeletons those of the princes? In view of the very precise position given by Sir Thomas More, it seems very likely. At the very least, it is a most uncanny coincidence. Dental evidence seems to support the idea that the skeletons were those of the princes. The older child's teeth suggested that he (or she) was aged between twelve years nine months and fourteen years five months. This fits in with the age of Edward V, who was thirteen years old in 1483, the year in which he and his brother were believed to have died. Moreover, the skeleton of this child showed that it had a congenital dental abnormality, for both upper second premolar teeth are absent; Edward suffered from pain in the jaw. Lady Anne Mowbray, a close relation of Edward, also had congenitally missing teeth.

It is important to remember that the Tower of London contains many skeletons. For example, two skeletons, similar to those found under the White Tower, were found walled up in an old chamber. These could equally well have been the skeletons of the princes. Another discovery was that of an Iron Age youth, found during excavations of the Inner Ward.

Two other points are worth considering. First, the depth at which the skeletons were found. It has been suggested that ten feet is quite deep, even for a secret grave, and that this depth suggests a much older burial. Yet, once again, Sir Thomas More was emphatic about the great depth of the grave. The second point is that, when the skeletons were first discovered in the seventeenth century, they were described as being found with "pieces of rag and velvet about them". Velvet was invented in Italy in the fifteenth century and was not manufactured in England until the sixteenth. (Velvet would have been imported into England in the fifteenth century.) Of course, it is quite possible that the person who gave this description used the word velvet to mean any kind of expensive cloth, but if the material was, indeed, velvet, then there is a strong probability that the remains are those of the Princes, not only because of the timing, but because only those of the very highest rank would have worn clothes made of this material.

With the modern techniques at the disposal of forensic scientists, many questions could be answered if a new examination of the remains took place. However, it seems unlikely that permission for a fresh investigation will be given. The bones are interred in an urn in Westminster Abbey.

Apart from being of intrinsic historical and scientific interest, this case shows how an investigation can be mishandled as a result of an incorrect approach from the outset. The "correct" answer was "known" before the forensic investigation began, with the result that the very question being asked could not be answered. This is not uncommon, as we shall see later in this chapter.

MARK OF THE BEAST

Bite marks inflicted on the victims of murder, especially women, leave a particularly unpleasant impression on the mind. Professor Keith Simpson, England's first Professor of Forensic Pathology, called it the "Mark of the Beast".

The silver lining of this unsavoury cloud is that the bite marks may betray the perpetrator. One of the first cases in which a conviction was brought about by bite-mark evidence began when a young couple, Mr and Mrs Gorringe, left a dance hall in Tunbridge Wells just before midnight on December 31, 1947. They were quarrelling as they left.

In the small hours of New Year's Day, 1948, Mrs Gorringe's dead body was found in a yard behind a lorry; she was still wearing her dance frock.

Detective Superintendent Frank Smeed, head of the Kent CID, took charge of the case and asked Professor Simpson to help him in the investigation. The woman's head had been battered in and she had been strangled; and, significantly, she had a bite mark on her right breast.

Wax impressions of the mark were taken, revealing a set of teeth that were so badly spaced and angled that they seemed quite distinctive. They were compared with the teeth of Mr Gorringe, the prime suspect; and they fitted exactly. Gorringe was put on trial, convicted and sentenced to death, although he was later reprieved.

Although the Gorringe case was a pioneering one, it was relatively straightforward. A subsequent case, in which Professor Simpson was involved, was more complex. On August 7, 1957, the body of Linda Peacock was found battered and strangled in a cemetery in Biggar, halfway between Edinburgh and Glasgow in Scotland. She was found by two policemen, lying between a tombstone and a yew tree. Again, she bore bite marks on her right breast.

Working in collaboration with Dr Warren Harvey of the Scottish Detective Training School, Professor Simpson took impressions of the marks. Meanwhile, the police interviewed three thousand people and excluded all but twenty-nine individuals from the investigation; they were all young men from a nearby detention centre. When asked, they all agreed to allow the police to take impressions of their teeth. One set of teeth, designated number fourteen, looked the most likely, but they didn't quite fit. Indeed, the marks left by the teeth looked decidedly odd. Number fourteen was, eventually, eliminated.

The investigators, convinced that one of the men from the detention centre was the murderer, were puzzled by the lack of a definite "fit" by any one of the sets of teeth. But the problem was solved in a most startling way. Professor Simpson and Dr Harvey had assumed that the assailant's head was oriented in the same way as the head of the girl, but, by turning the cast impression round through an angle of 180°, the teeth of number eleven fitted perfectly; the man had bitten his victim by approaching her from behind – the heads had been upside down in relation to one another.

Gordon Hay, a 17-year-old youth, was number eleven. He was known to have met Linda and had been with her the day before the murder. Dr Harvey wanted to make absolutely certain of the identification. Hay's teeth had strange crater-shaped pits on his canines; and Dr Harvey wanted to see how common this condition was. He examined one thousand canines in 342 boys aged sixteen or seventeen and found only two with pits resembling those on Hay's teeth. Moreover, not one of the boys had more than one such pit, whereas Hay had several. He was found guilty of murder, but, because of his age, was sentenced to be detained at Her Majesty's pleasure.

CRANIO-FACIAL RECONSTRUCTION

One of the techniques of human identification that has become an indispensible tool in forensic identification is cranio-facial reconstruction. This is a technique that produces an actual, three-dimensional model of the head and face from a skull. It is a very complex technique, requiring both the strict adherence to scientific facts, as well as an artistic flair in producing an actual model that people can look at and, hopefully, recognise.

While the application of this technique requires great skill and experience, the basic idea is simple enough. The muscles of the face "reflect" the shape of the skull beneath them; in other words, the thickness of the flesh at any given point will be largely dictated by the shape of the bone on which it lies. When a cranio-facial reconstructionist sets to work, he fixes little pegs at various key points on the skull, the height of the pegs being equal to the calculated thickness of the flesh at that point. Layers of clay, simulating the muscles, are then placed over the pegs and eventually a face appears.

People who do this kind of work are sometimes surprised by the kind of face that emerges. Others, such as Dr William Aulsebrook, keep the skull at home for a time, doing nothing whatsoever about it, apart from "getting to know it". In fact, the actual skull is not used in the reconstruction, but a cast made from the original; this is because the skull itself may provide other kinds of evidence that would disappear under a layer of clay. Advances in clinical imaging and computer technology have led to the development of a number of computerised cranio-facial reconstruction systems that differ in the level of automation, the amount of input required from the operator and the reliability of the reconstruction.

Another computerised technique that is used for human identification is that of video-superimposition. This grew out of photographic superimposition, which was simply the laying of a radiograph of a skull over a photograph to see whether the two would fit. Video-superimposition is essentially the same technique, but here the images can be manipulated by computer. This is now one of the most commonly used methods for confirming identity, especially in forensic cases. However, there is concern about the ease with which skulls and photographs can be made to match; and the exponents of this technique are themselves aware of this problem and advocate further research.

However, cranio-facial reconstruction is still the method of choice when the identity is completely unknown, however sometimes, it can be difficult to make a cast to work from. When reconstructing the face of an Ancient Egyptian mummy, Aulsebrook and his co-workers took sectioned images of the skull as it lay within its wrappings. This was done using a technique called cat-scanning, in which the skull is "sliced" into a number of pieces that can be rebuilt on a computer. In this way, a three-dimensional computer image was produced. It was then used to make an actual model of the skull, which was used to produce the cranio-facial reconstruction. The resulting face of the priest, Peten Amen, was uncanny to see.

Britain's foremost exponent of cranio-facial reconstruction was Richard Neave of the University of Manchester (now retired). For many years he produced "miracles", as one admirer put it. One of his most famous cases was the reconstruction of the head and face of King Philip of Macedon, the father of Alexander the Great. An interesting point about this case is the fact that Philip received a dreadful battle wound on the forehead and right eye, features that are visible in some of the contemporary images made of him. We have seen above how anthropology can shed light on the medical history of a skeleton; and Philip is a case in point. The skull showed that the king had, indeed, received a horrendous wound in the right place; in fact, it is quite astonishing that he survived such a wound at that age in history, since, in the absence of refined surgical techniques and antibiotics, most people would have died after being severely injured in this way.

Much of the efforts of cranio-facial reconstructionists have been directed toward the production of faces of historical characters, whose physical appearance was unknown. Examples of such people are Johann Sebastian Bach, whose face was reconstructed late in the nineteenth century. The twentieth century saw the reconstruction of the face of Czar Ivan the Terrible by the Russian Gerasimov, while in Britain the face of King Midas was reconstructed by Neave. Inevitably, there have been doubts about the accuracy of the finished face, since there was no way in which it could be compared to the original face, or to a known visual representation of it. It was the application of the technique to forensic science that placed cranio-facial reconstruction on a firm footing, since now the faces could be checked against originals. In fact, independent studies on the reconstruction of faces from skulls whose "owners" appearance was known have shown that the technique is generally very successful, although the application to forensic science was the acid test.

A case example will make this point clear. In March, 1973, a young man named Johann Boucher disappeared after a car accident near Ladysmith, South Africa. No trace of him was found, although a picture of him was published in the local newspaper, *Die Nataller*. Boucher had been a storeman working for the South African Railways, but his employers and work-mates were unable to shed any light on his likely whereabouts.

In 1987 staff of the Durban Department of Parks, Recreation and Beaches found a skeleton lying under shrubs in the vicinity of Burman Bush. The skeleton remained unidentified for two years, all attempts at matching dental records with the teeth having failed. Eventually, the police approached Dr Aulsebrook, who proceeded to reconstruct the skull's features. It was only after he had done this that the police showed him a photograph of Johann Boucher – the resemblance between the two was remarkable.

A newspaper, the *Mercury*, published a picture of Aulsebrook's reconstruction of Boucher's face; shortly after that the newspaper received a telephone call from a man who said that he recognized the face as being that of Boucher.

IN THE BLOOD

Blood, of course, is one of the commonest things found at the scene of a violent crime. Almost everyone knows about blood-grouping as a method of narrowing down the number of suspects, but what is not generally known is that the famous ABO system is only one of fourteen methods of blood-typing. However, ABO is probably the most commonly used system. All systems are based on the types of antigen (a protein molecule capable of binding on to an antibody) on the red blood cell's membrane.

In the ABO system the two antigens are known as A and B; and four blood-groups are determined according to this system – A, B, AB and O. People having the blood-group A have the A antigen; those belonging to the B group have the B antigen; those belonging to the AB category have both antigens; while those classified as belonging to the O group have neither antigen.

The antibodies present in a person's blood correspond to the opposite group, thus A-group people have b antibodies; those having B antigens have a antibodies; AB people have neither a nor b antibodies; and, finally, O people have both a and b antibodies. If, say, b antibodies are introduced in to the blood of a B-group person, the reaction will make the red cells clump together, resulting in death. The test used for determining the blood-group of a person thus depends on the addition of antibodies to a sample of blood and noting the reaction. It can easily be seen that blood-group O can only give a negative result, although other, more specialized methods are used to determine blood-group positively. Contrary to popular belief, the antigens used in ABO testing are present, in most people, in other body fluids, such as saliva and sweat, so that even traces of saliva on a piece of cloth may reveal a person's blood-group.

DNA FINGERPRINTING

We come now to that almost magical technique known as DNA fingerprinting. This technique is seen as being the most powerful and reliable tool in the forensic scientist's human identification armoury. Yet it is a much misunderstood technique, as we shall see.

DNA is the genetic material of the cell. It is what largely determines our physical characteristics, although much depends on the way in which environmental factors affect the way genes are expressed in the individual. This point need not concern us further here, but we shall return to it in the last chapter of this book.

DNA, short for deoxyribonucleic acid, is present in the cell nucleus and in extra-nuclear organelles of the cell, known as mitochondria (singular: mitochondrion). We inherit half of our nuclear DNA from our fathers and half from our mothers, but we receive all our mitochondrial DNA (known as mtDNA) from our mothers. Thus, nuclear DNA can yield information about our paternal and maternal relations, while mtDNA can shed light only on our matrilineal descent.

The basic idea is that certain stretches of DNA are believed to be unique to an individual – no-one else will have the same DNA along those particular stretches, unless they were the identical twin of the individual in question. When a sample of tissue – be it blood, semen, skin, etc. – is found at the scene, it can be collected and used as a source of DNA. Once the DNA is extracted it is cut into little pieces by certain enzymes that act as chemical scissors. This mixture of DNA pieces is placed on a plate in a gel through which an electric current is passed. The different pieces will separate from one another as they move along the gel, the larger ones will move faster and farther than the smaller ones. The result is still invisible to the human eye, so radioactively labelled pieces of DNA are added; these adhere to the pieces already separated. The radioactivity makes the pieces visible when an X-ray film of the gel is made. In this way a "picture" of the DNA can be made and compared with a similarly-prepared "picture" of the DNA from the suspect. If the samples match exactly, one has a positive identification. The now-famous DNA profile is this picture, which is a series of bands resembling the bar code on supermarket products.

An example of the efficacy of this undoubtedly powerful technique took place on February 21, 1987, in Orlando, Florida, USA, when a woman, asleep at home, was viciously beaten and repeatedly raped by an intruder. Later, the police found two fingerprints on the window-sill. They then arranged for a swab to be taken from the woman in order to obtain a sample of the assailant's semen.

In March, the police received a call saying that a prowler had been seen in a particular part of the city; a police car that went to the area in response to the call saw a car speeding away. The police car followed it, until their quarry turned round a sharp corner and crashed his car. The driver was arrested and identified as one Tommie Lee Andrews. Another woman, the victim of an earlier rape, identified Andrews as her attacker. He was put on trial for rape, battery and armed burglary. The fingerprints from the window-sill proved to be those of Andrews. Further tests were conducted and it was shown that the semen belonged to a person with blood-group O; Andrews was blood-group O. Finally, DNA tests on the semen, Andrews' blood and the victim's blood were compared. The results showed clearly that the semen came from Andrews. DNA from semen recovered from the victim of the earlier assault also showed it to be the same as Andrews's DNA. Tommie Lee Andrews was found guilty as charged and sentenced to one hundred years in prison.

The perceived infallibility of DNA evidence has had a remarkable effect, not only on police officers eager to believe that they had a guaranteed method of identification, but also on criminals themselves. In a British case of murder and rape, the criminal in question paid another man to impersonate him when the police were taking DNA samples from people living in the area. Happily, the deception was discovered and the man was arrested and convicted.

DNA fingerprinting has had many other notable successes; its failures, on the other hand, have not always received similar publicity.

"*Get me every 'ologist under the Sun!*" These were the words of Detective Chief Superintendent John Williams, the officer in charge of the case of Karen Price, the girl whose skeleton was found wrapped in a carpet and buried in the garden of a derelict house in Cardiff in 1989.

John Williams's order was obeyed by his staff, with the result that the investigation of the murder of Karen Price became a classic of forensic detection. I was involved in the case, but the star of the show was Richard Neave, who reconstructed the girl's face, with the result that she was recognized and identified by several people. DNA samples taken from the bones and compared with Karen's parents confirmed the identification.

I applaud Williams's approach to the task, as expressed in his instruction to his staff, not only because it displays a thoroughness and a desire to leave no stone unturned, but mainly because it demonstrated an understanding of the nature of evidence. I am sometimes asked why we need to bother with techniques such as blood-typing, cranio-facial reconstruction, fingerprinting (the old kind), anthropology and all such methods, when we now have the highly scientific and infallible DNA testing.

The answer, of course, is that, until recently DNA fingerprinting is of absolutely no use, unless one had some idea as to whom the victim or the culprit may have been. In the Karen Price case, if DNA had been extracted and a profile made, no-one would have been able to make any sense of it. People do not recognise DNA profiles as things they have seen before. Only when Neave came up with the face – an idea or a hypothesis that could be tested – did DNA come into its own.

Today, a mere twenty or so years on from the Karen Price case, the situation is somewhat different. In 1998, Britain established the World's first national DNA database (NDNAD). DNA is taken and a profile is recorded for every person arrested for a recordable offence. Many other countries, most notably the United States, Canada, Australia and many European countries have followed suit. It should be noted that, in all cases, the profile is composed of only a small part of the total DNA profile of an individual, but the areas of DNA chosen were those that discriminate widely between individuals and so far there have been no reports of two unrelated people sharing the same profile. Improvements in the techniques for both extracting and amplifing (making more copies) the smallest traces of DNA mean that it is now possible to obtain a DNA profile from a sample as small as that found in the saliva left after licking the back of a stamp. This profile can be fed into the NDNAD and matches searched for, sometimes with spectacular results. In June 1989, Peter and Gwenda Dixon failed to return to their home in Oxfordshire at the end of their annual holiday in Little Haven, Pembrokeshire. Their bodies were found a few days later in undergrowth a mere twenty yards off the coastal path not far from the camp site at which they had been staying. They had both suffered shotgun wounds. Despite knocking on every door within a ten mile radius and tracking down potential witnesses across the length and breadth of Britain and Europe, police found no culprit, until a cold case review was launched in 2006. The material collected in 1998 was re-examined and, amongst other things, DNA on a pair of shorts linked a local Milford man and convicted burglar, John Cooper, to the Dixon's murder. In May 2011, he was tried and convicted of both the Dixons' murder and the earlier murder, in 1985, of a local brother and sister, Richard and Helen Thomas, although he has launched an appeal.

However, no evidence is infallible – not even DNA or old-fashioned fingerprints. Arriving at the truth in a forensic investigation is a matter of attacking the problem from many different angles. If the story that emerges is supported by all the strands of evidence, one can then have good reason to believe that it is the truth.

Let us now look at a case that includes many strands of evidence, a story, moreover, that has received a great deal of international attention. Yet, despite all the work that has been carried out to resolve the problem, it remains as far from resolution as ever.

THE ROMANOV CASE

On May 10, 1992, the world woke up to the news that the mortal remains of Czar Nicholas II, Czarina Alexandra, their five children and several retainers had been discovered in a pit in the Koptyaki Forest, near Ekaterinburg, where the massacre of the Imperial Family was alleged to have taken place. Specifically, it was claimed that the skeletons of all family members were found. After a number of forensic investigations, including DNA tests, were carried out, the remains were buried at a ceremony in St Petersburg on July 17, 1998, eighty years to the day after the disappearance of the Romanovs.

Despite the fact that the remains were described as being unquestionably those of the Czar and his family, the burial was boycotted by Alexi II, Patriarch of the Russian Orthodox Church, and by several members of the Russian nobility, including some members of the Romanov family. The Czar's nephew, Tikhon Kulikovsky, described the whole affair as a "tourist stunt to make money" and denounced the bones as false. President Boris Yeltsin, too, had announced that he would not attend the ceremony, but he changed his mind at the last minute, although he referred to the bodies as the remains of those of innocent victims, not as the Czar and his family. Metropolitan Ioann of St Petersburg spoke of his disquiet over the "secret means" of the discovery of the remains and said that "there has been an attempt to fool the public with these bones". If science had established the identity of the remains beyond reasonable doubt, as people were led to believe, why did some of those individuals most closely associated with this issue refuse to accept that the remains were genuine?

The realization that something was wrong came soon after the discovery of the remains was made public. The initial reports by the discoverer, Geli Ryabov, stated that eleven bodies had been found, but the photographic evidence showed only nine. Ryabov had been a member of the KGB and could not have acted without information from that source. He and his associate, Alexander Avdonin, had been given permission to work in the Central State Archive at the height of the Soviet era, suggesting that the Soviet government was involved from the outset. In 1979, the year in which the discovery was made, Boris Yeltsin was Communist Party boss at Ekaterinburg. Yeltsin had ordered the demolition of the Ipatiev House, the site of the imprisonment and alleged massacre of the Romanovs, two years earlier, ostensibly to make room for a car park. More interestingly, a detailed examination of the site in the Koptyaki Forest by the White Russian Commission of Nikolai Sokolov in 1918 failed to find anything at that spot. Subsequent searches had similarly negative results, although the area was the haunt of souvenir hunters every Sunday. The historian Vladimir Bolshakov remarked that a group of people had been seen tampering with the site about one month before its discovery by Ryabov and Avdonin.

Acting without consultation with Moscow, the regional government of Ekaterinburg invited a team of American forensic specialists, led by Dr William Maples of the C.A. Pound Human Identification Laboratory at the University of Florida in Gainesville, to examine and report on the remains. Maples immediately made it clear that only nine skeletons had been found; the Czarevitch and one of the daughters were missing. Although the Russian authorities insisted that the missing girl was Grand Duchess Maria, Maples was equally certain that it was Anastasia, the youngest daughter. Anastasia was just seventeen at the time of the disappearance of the Romanovs and Maples showed that the skeletons of the three girls in the pit had all their cranial bones fused, whereas this would not have been the case if one of them had been Anastasia. The Russian assertion that the missing skull was Maria's was based on a computerized reconstruction of the faces of the skulls by the Russian scientist Sergei Abramov, although the facial bones of the skeletons had been very badly smashed, to the extent that Maples said that he "could have created George Bush or Bill Clinton from the same skull". Professor Viacheslav Popov of the St Petersburg Military Medical Academy said that all the cardinal points upon which identification could be made were absent from the smashed-up skulls. Despite intense opposition, both from the Russian authorities and the press, Maples and his team insisted that, if the remains were, indeed, those of the Romanovs, the missing girl must be Anastasia.

Maples then raised the possibility that the bodies may not have been buried in 1918, but later. If death and burial had taken place in 1920, the bones in Anastasia's skull would have been fused. Interestingly, despite this observation and the fact that the site had been searched before with negative results, no proper investigation into the age of the pit was carried out in the early stages. An almost unpublicized investigation of the age of the pit was carried out by the Russian authorities; it appears to have concluded that the grave was between fifty and sixty years old – in other words, it was placed firmly in the 1930s, at the height of the Stalinist terror. I have not seen the original report and I do not know what methods were used. In 1994 I offered to carry out such an investigation on the pit and the soil that had been removed from it, and to bring together a team for that purpose. My offer was ignored by the Russian authorities. When I contacted the Russian Embassy in London and asked to speak to the Cultural Attaché about this matter, I was told very firmly that no such official existed. All my letters to forensic scientists in Russia were ignored after that date, in spite of the fact that a delegation of Russian forensic scientists came to see me in Cambridge in 1991 in order to discuss the applicability of my own specialism to, among other things, the investigation of burial sites.

Another odd fact came to light during the Ekaterinburg investigation. Although the Romanovs were said to have been shot and bayoneted to death, there were no scars on the bones. This does not necessarily mean that the remains do not belong to the Romanovs, but it does seem to suggest that the accepted story of how they were killed is false.

There were questions, too, over the positions of the skeletons in the pit. Speaking at a public session of the Duma on May 21, 1998, Professor Popov demonstrated that the skulls of skeletons five and six could not have been found at the north-east corner of the pit, as stated by Ryabov and Avdonin, who were unable to answer the State Prosecutor on this point.

More worrying still is the authenticity of some of the archival material recently released by the Kremlin. The document known as the "Yurovsky Note", allegedly the account of events written by Yakov Yurovsky, the man in charge of the Ipatiev House at the time of the execution of the Imperial Family, has been shown to be a forgery produced by Stalin's master forger Pokrovsky, who once announced that "history is politics". The note was written in his own hand on paper that could not have been manufactured before 1924. My own letters to the Russian Archives, asking whether certain early documents had survived, were ignored; my approach to the London embassy in this matter met with a response similar to the one I received over the matter of the proposed pit investigation. Such a response is meaningless if the papers had not survived, for then the simple response would have been that the papers cannot be found; the wall of silence that descended can only be interpreted as an attempt to conceal evidence.

No foreign investigator, including Maples and his team, ever saw the bones in situ. When the American specialists arrived, all the bones had been removed from the pit, washed and placed, uncovered, in the Ekaterinburg mortuary. The remains had been handled by a number of people, including American Secretary of State James Baker. Over one thousand bones were missing from the nine bodies – more than half the total. By way of example, half the vertebral bones of the skeleton identified as belonging to Grand Duchess Tatiana were missing. This suggests that the bones had been moved from one burial place to another during their history. There was no proper chain of custody, as is customary in forensic investigations. When samples of the bones were brought to London by Dr Pavel Ivanov for DNA testing, he carried them in his Aeroflot flight bag, together with his sandwiches.

The mtDNA work carried out by Dr Peter Gill and his colleagues appeared to show that the skeletons believed to be those of the Czarina and her daughters had been correctly identified. However, these identifications were based on a comparison with a blood sample donated by the Duke of Edinburgh alone; no other relations of the Czarina and her daughters were tested for further comparisons. In view of the subsequent disquiet over the reliability of mtDNA as an identification tool (as we shall see below), comparisons with other relations would seem to be indicated.

However, problems did arise with the mtDNA from skeleton number four, identified as being that of the Czar. Samples from that skeleton were compared with samples from two living relations of the Czar, Countess Xenia Cheremeteev-Sfiris and the Duke of Fife, but, although similar, the samples did not match at position 16169. This was interpreted as a heteroplasmy, or mutation. In order to look into this problem further, the body of the Czar's brother, Grand Duke George, was exhumed and mtDNA samples taken. These showed that the Grand Duke had a heteroplasmy at the same position and it was concluded that this was an exact match. This is not quite true; although there is a heteroplasmy at the same position, it is not the same heteroplasmy. At position 16169 the DNA of skeleton four was seventy-two per cent cytosine and twenty-eight per cent thymine, whereas at the same position on Grand Duke George's DNA it was almost the reverse, being seventy per cent thymine and about thirty per cent cytosine. George's heteroplasmy matches those of Countess Cheremeteev-Sfiris and the Duke of Fife; skeleton number four does not.

It seems to me that the reasoning over the use of DNA technology to determine the identity of skeleton four came perilously close to saying that, if the results matched, the skeleton was the Czar, but if they did not match, it was still the Czar. It occurs to me that the results would have been more compelling if the tests had been conducted "blind"; in the event, the expected "correct" answer was given before the tests were carried out. In saying this I do not mean to impugn the integrity of any of the scientists involved, but I do say that, once a preconceived idea is planted in the mind, it is very difficult to ignore it. In this matter, I speak from experience in forensic science case-work; I always ask not to be told the conclusions of other workers in the case, until I have conducted my own examination and reached my own conclusions, uninfluenced by the conclusions or hypotheses of others.

It is highly unlikely that skeleton four is that of the Czar. When Nicholas, as Czarevitch, visited Otsu, Japan, in 1892, he was attacked by a police officer who thought that he was desecrating a sacred monument. Nicholas received several blows to the head and his clothes were drenched in blood. A piece of bone was dislodged from his forehead and he bore the visible mark for the rest of his life.

No such mark or bone damage appears on skull number four. When the height of skeleton four was first measured by Russian scientists it was estimated that the living person had been six feet tall. When Maples examined the bones, he found that the femurs and long bones of the arm had been sawn through and shortened. When he asked when and why this was done, he was given no explanation. His measurements, based on the new bone lengths, produced a height estimate of five feet and seven inches – the exact height of the Czar, who was a short man.

Interestingly, Grand Duke Michael, the Czar's other brother, who disappeared in 1918 and whose body has never been found, was six feet tall. Even before the alteration in the lengths of the bones, the arm bones were disproportionately long – long enough to make the living individual "look like an orang-utan" in the words of William Maples, who postulated that the arms of one of the servants, Trupp, may have been mixed up with other bones to form skeleton four. The possibility that some of the bones belonged to Grand Duke Michael seems very real. It is also clear that, if the bones are those of the Grand Duke, his remains could appear to be those of the father of the three daughters, without actually being so.

Another problem arose with the DNA tests. A blood sample from Princess Sophie of Hanover, the Duke of Edinburgh's sister, was sent to the University of California at Berkeley, but the results do not appear to have been published. Maples, who arranged for the work to be done in Berkeley, was not given an answer and he surmised that the sample may not have matched the sample from Prince Philip. This increased his disquiet over the reliability of mtDNA as an identification tool, since he knew of two children who were undoubtedly the offspring of the same mother but whose mtDNA did not match. It is also interesting that each of two matrilineal descendants of Queen Victoria, Princess Charlotte and Princess Feodora – a mother and daughter – have mtDNA that does not match the other.

There have been several claimants to the identity of the missing Romanovs or their descendants. What can forensic science say about them and the validity of their claims? The most famous claimant, known for most of her life as Anna Anderson, was believed by many people to have been Grand Duchess Anastasia. Recent DNA studies have purported to show that she was not whom she claimed to be, yet it now appears that the tissue (a slide-mounted polyp) from which the DNA was extracted had been modified and should not have been used in the tests. Setting aside for the moment the forensic science evidence supporting her claim, her memories, her voice, her mannerisms and her "presence" convinced many people that she was whom she claimed to be. There is also some historical evidence from eyewitnesses attesting to the fact that she escaped the massacre. During one of the trials to establish her identity, the documentary evidence in support of her claim had to be wheeled into the court-room on forty-nine trolleys, more evidence than any other person is likely to be able to adduce in support of their identity.

The story of Anna Anderson is very well known, so I will not go over it in detail here. However, the forensic evidence is worth rehearsing briefly. Anna Anderson and Anastasia shared a number of similar physical characteristics: both had a *hallux valgus* at the base of the big toe in both feet and in both the condition was worse in the right foot; both had middle fingers that were much shorter than normal; they had moles in the same places; they had bayonet wounds of a shape that could have been made only by a certain make of Russian bayonet; in both of them the region between the lower lip and the chin was more or less flat, not concave, as it is in most people; photographic analyses of the faces of the two women produced the conclusion that they were the same person; four detailed studies of the ears gave the same conclusion. More subjectively, several people who knew Anastasia before 1918 commented that Anna Anderson had the same voice as the Grand Duchess and that she simply looked like Anastasia. On the other hand, several others said that she did not resemble the claimant.

Although such things as memories may be regarded as being too subjective for scientific analysis, the totality of Anna Anderson's memories must provide evidence for identity and could, in principle if not in practice, be analysable using scientific techniques. Anna Anderson remembered so many things that could only be known by members of the Imperial Family or one of their close circle and it is valid to ask what is the probability of someone who was not such a person knowing all these things. The answer must be "very small indeed", although it may be well-nigh impossible to put a figure on it. If one added the probability of having such recollections to the probability of having all the distinctive physical attributes, the overall probability that Anna Anderson was not Anastasia must be even lower. Interestingly, the paper that concluded that the claimant was not Anastasia included a discussion of probability, but did not use most of the available evidence in calculating that probability, concerning itself mainly with the DNA evidence.

One author, whose name it would be kinder not to mention, argued that the scientific evidence (by which he meant the DNA) pointed in one direction and that the "other" evidence pointed in another direction. The author, who is not a scientist, concluded that the scientific evidence should be believed and the rest of the evidence should be ignored. This reveals a total misunderstanding of the nature of science.

Contrary to some very widespread beliefs, science is an endeavour that is concerned mainly with probabilities, not absolutes. Scientists do not "know" things in a different kind of way than other people know other things, they can only suggest a probability for something happening, whether it be the exact position of an electron at a given moment or the behaviour of an antelope when it sees a lion as we saw in Chapter 1. DNA results are no exception to this rule; they are no more "scientific", no more "true", than the rest of the evidence in the case. They are as much subject to the laws of probability as anything else. In science, one accepts an idea (a concept, a hypothesis or what have you) as long as there is no evidence that casts doubt upon it. The evidence in support of Anna Anderson's claim to her identity not only casts doubt upon the DNA results, it overwhelms them.

Due to its many impressive results and its powerful scientific base, DNA evidence has acquired a kind of invincibility that cannot and should not be accorded to any scientific discipline. It is as though a probability of "X" using DNA results is more powerful than the same probability using other evidence. This cannot be supported logically, but such is the strength of belief in DNA evidence that its results often appear to make many people, including some scientists, override logic. It is extremely dangerous to give a particular scientific technique such an aura of infallibility, as this tends to close the mind against other valid evidence. Nevertheless, so many people tend to have a strong attachment to particular lines of evidence based on certain currently popular techniques. This cannot be justified.

There have been many other Romanov pretenders or claimants to Romanov descent. Most of these claims are clearly fraudulent and can be dismissed. However, there is one recent claimant whose case is so compelling it simply cannot be ignored. Dr William Lloyd-Lavery is a historian living in Ireland who has undertaken many years of painstaking research into the issue of Romanov survival. I was surprised to find that, although a large amount of scientific, photographic, testimonial and circumstantial evidence exists, this case has gone largely unnoticed.

I will not rehearse all the documentary and other evidence supporting Lloyd-Lavery's claim, although there is a great abundance of it. I will restrict myself to the scientific evidence. Two conclusions were reached: first, that the Czarevitch was not murdered in 1918, but survived, living under the assumed identity of Nicholas Chebotarev and dying only in 1987 at the age of 82; and, secondly, that Dr Lloyd-Lavery is his son. I will take each conclusion in turn.

It is often remarked that, even if the Czarevitch had not been murdered in 1918, it is highly unlikely that he would have survived for very long, since he was a haemophiliac. In spite of what has been said repeatedly, there is very little evidence that the Czarevitch suffered from this disease. Queen Victoria, who is said to have carried the gene for haemophilia and to have transmitted it to her descendants, had no doubt about the truth: "This disease is not in our family," she said. It certainly would appear that this is the case, since exhaustive searching of Queen Victoria's ancestry over a period of eight generations by Professor William Bullock and Sir Paul Fildes in preparation for their book on haemophilia produced not a single case of the disease. However, there are plenty of examples of another disorder, the "royal malady" of porphyria.

All Queen Victoria's allegedly haemophiliac descendants died of other causes, such as accidents, or lived to a normal age. Nevertheless, they did suffer from some kind of bleeding condition, although the available evidence does not support the assertion that that condition was Factor VIII Haemophilia A, as has been stated over the years. At the time of the Czarevitch's illness, many very different conditions were referred to collectively as "haemophilia", although the state of medical knowledge at that time was such that they could not be distinguished from one another. No official record states that the Czarevitch had haemophilia. Far from being a sickly child, the Czarevitch was, in fact, a strong and healthy boy. The belief that he had haemophilia springs from a comment made by Princess Catherine Radziwill, who said that the boy "suffers from an organic disease of the arteries which are liable to rupture upon the slightest provocation and even without cause". All subsequent remarks about the Czarevitch's haemophilia stem from that comment. Princess Radziwill had an intense hatred for the Romanovs and her statement can arguably be put down to malice; several other comments she made about the Romanovs are demonstrably untrue.

The probability is that the Czarevitch suffered from another, non-life-threatening, bleeding condition. Thrombocytopenia is a possibility. Porphyria, the disease that caused the "madness" of King George III, an ancestor of the Czarevitch, is also possible. Factor IX Haemophilia B, known as Christmas Disease, is yet another possibility. The first sign of bleeding in the Czarevitch occurred when he was a very small baby; he bled from the navel. This condition is not rare among babies; my own son had it and he is not a haemophiliac, nor does he suffer from any other bleeding disorder.

All the available evidence on the Czarevitch's condition was submitted to Dr Rory O'Donnell, a consultant haematologist, who reported that "Factor VIII Haemphilia A is extremely unlikely. The hall mark of this condition, normally confined to males, is serious joint bleeding, which was never evident. Epistaxis and GI [gastrointestinal] bleeding [which were experienced by the Czarevitch] would not be found in Haemophilia. There are very many congenital and acquired forms of bleeding disease and none of these would have been recognised at the turn of the century." He went on to say that "thrombocytopenia is a possibility" and that "systemic lupus erythematosis, which may have bleeding manifestations, or indeed porphyria are also possible". He also wrote that the Czarevitch "could have lived to a great age".

In 1995 Professors William and Malcolm Potts argued that either Queen Victoria'a haemophilia gene was a mutation, or that she was illegitimate, the Duke of Kent not being her real father. However, if "this disease is not" in Queen Victoria's family, as the new evidence appears to suggest, then neither hypothesis is necessary and the mystery disappears.

So, the Czarevitch could have survived, but did he? Nicholas Chebotarev lived in Yugoslavia, the USA, Ireland, England and, for a period, in France. Throughout his life he was surrounded by rumours that he was a prince and, specifically, that he was the Czarevitch. His friends called him Alexei, which was the Czarevitch's name. Dr William Lloyd-Lavery is has documented in some detail, in his 1998 book *Blood Relative*, the evidence supporting his belief that Chebotarev was, indeed, the Czarevitch. It is a long and complex detective story. The evidence is compelling, but, being mainly of a historical and documentary nature, I will not go over it here. Suffice to say that new scientific investigations, currently in progress, have given support to the belief that Chebotarev was the Czarevitch.

However, there already exists in the public domain scientific evidence that supports the claim that William Lloyd-Lavery is related to the skeleton identified as belonging to the Czarina and to skeleton four in the Koptyaki pit. DNA work carried out by Dr Gill gave results that support the belief that the skeletons could have been Lloyd-Lavery's grandparents.

This identification was supported by a cephalometric analysis conducted by Professor Popov in St Petersburg on X-rays of the skulls of William Lloyd-Lavery and the Czarina. After comparing thirty-eight key measurements of Lloyd-Lavery's skull with those of the Czarina and her three daughters (skeletons three, five, six and seven in the burial pit), Popov concluded that Lloyd-Lavery matched the skulls in ninety-eight per cent of cases. Popov was also able to show that Lloyd-Lavery's skull and that of the Czarina shared four features in common: a protrusion at the back of an elongated skull; the presence of Wormian bones in that region; a horseshoe-shaped lower jaw; and overlapping teeth in the lower jaw. Each one of these features is present in about ten per cent of the population. Lloyd-Lavery's blood-group is A, the same blood-group as the Czarina.

It seems clear that there is scope for further forensic work to investigate the fate of the Romanovs. It is highly regrettable that the Koptyaki bones were buried so hurriedly and after an investigation that was, in certain respects, flawed. However, certain facts have now emerged; and they need to be better and more widely known. The Czarevitch's skeleton is missing, as is the skeleton of one of his sisters;

despite exhaustive searches no trace of them has been found. It is a reasonable assumption that they survived the massacre. Clear evidence that the Czarevitch did not suffer from Factor VIII Haemophilia A and that he could have lived to old age is available. There are good forensic grounds for believing that skeleton four is not that of the Czar. DNA and osteological (i.e. bone) data have linked Dr Lloyd-Lavery to the Czarina and her daughters. This evidence prompted William Maples to say that the probability that Lloyd-Lavery was the grandson of the Czar is "a serious proposition". The solution to the Romanov mystery may be within reach.

As for the accepted story of the fate of the Romanovs, it is salutary to remember Napoleon's rhetorical question: "What is history, but a fable that has been agreed upon?"

* This is in line with the conclusion reached in 1973 by the International Association for Identification, after a three year study, that "no valid basis exists for requiring a predetermined minimum number of friction ridge characteristics which must be present in two impressions in order to establish positive identification." This view was upheld at a conference of members of the international fingerprint community in 1995 with the issuing of the Ne'urim Declaration.

** Advances in computer technology from the 1960s onwards ultimately led to the creation and development of Automated Fingerprint Identification Systems (AFISs). In the UK, Gerald Lambourne, head of Scotland Yard's Fingerprint Bureau is credited as being the first to work on methods of computerising the two million sets of prints held at that time. The UK's national fingerprint database (IDENT1) stood at 18.6 million sets of ten-prints as of April 2010. The first computer search of fingerprint files in the USA took place in 1980. The FBI's Integrated Automated Fingerprint Identification System (IAFIS) became fully operational in 1999 and links state AFIS computers to the FBI database, which currently contains records from over 70 million subjects or 700 million sets of ten-prints.

*** By using light of different wavelengths, ALS devices can be used to detect different kinds of forensic evidence, such as blood and semen as well as fingerprints.

****This remark does not apply to examiners in the United States, where there is more concern with professional standards than there is with the counting of matching points.

CHAPTER FIVE

CAUSES

Everything in nature is a cause from which flows some effect.
Baruch Spinoza

I was standing in the mortuary looking down at the dead body of a young man; he had been fished out of the Regent's Canal in London some time earlier. I was bending over to examine his chest more closely, when the police inspector, unable to contain himself any longer, said excitedly:

"There are twenty-eight of them, doctor! Count them!"

I counted: yes, there were twenty-eight stabs wounds through the chest, many of which went through the heart. They were not great gashes, but neat round holes. What kind of instrument could have made such holes, I wondered? A long, sharp, narrow, stiletto-like weapon, certainly; but what?

I stood up straight and the inspector, anticipating my question, said with a smile on his face:

"A kebab skewer, doctor! Frenzied attack!"

Both the man and his assailant had been drug addicts, and the killing had been done while under the influence of drugs. To stab someone twenty-eight times is by no means easy; anyone who doubts this assertion might like to make a stabbing motion with the hand twenty-eight times. It should soon be clear that such an attack, aimed at the same part of the body, can only be committed by an unhinged mind.

We enter now into the realm of causes – the "How?" of forensic science. In fact, it can be said that there are three causes in a crime, two of which answer the question "How?" and one of which answers the question "Why?" We shall leave the "Why?" question to the last chapter of this book; here we will consider the two "How?" questions.

Forensic pathologists distinguish between the *cause* of death and the *manner* of death. The cause of death can be something like heart failure, but the manner of death may be a stabbing through the heart. In other words, the manner of death is the cause of the cause. Another expression used in describing violent death is the mode of death. This is the way the person died, looked at from a different point of view – was it a "natural" death, was it suicide, or was it homicide?

The meanings of some of these words need further clarification. A "natural" death in the context of violence may mean, to use the classic phrase, that someone jumped and was not pushed. The injuries may well be the same whether the victim jumped or was pushed, but in the first case no human agency will have been involved; in the second a human being will have started the chain of events.

Earlier, I used the word "homicide" – not murder – advisedly, since "murder" is a legal term, carrying with it implications of intent – malice aforethought – but "homicide" is simply a descriptive epithet; it simply means the killing of one person by another – it could be murder, but it could also be manslaughter, excusable killing (e.g. self-defence) and so on. It is, therefore, the more general term.

STAB WOUNDS

Stabbing someone to death with a sharp-tipped weapon has a long history and is still a very common means of murder. The injuries caused by stabbing often appear so trivial, since a powerful stab with a sharp knife will enter the body with minimal damage to the skin. Moreover, a thrown knife will enter a body with greater force than will a knife held in the hand and stabbed into the victim.

A case in point concerned a young woman, who was cutting bread at a table. Her sister, sitting on a settee with her fiancé, was teasing her about her lack of success with boyfriends. Annoyed, the woman threw the knife at her sister, who turned away. The knife entered her head behind her left ear. Although her fiancé immediately pulled the knife out, she fainted and died about forty minutes later.

The post mortem examination showed such a trivial external wound that it was difficult to see how this could have caused death. In fact, the knife had penetrated under the base of the skull, passed through the jugular vein, through the neck muscles between the first and second cervical vertebrae, the tip coming to rest at the very centre of the lower medulla, or brain stem. Throwing a knife, even in jest, is very dangerous.

Stab wounds, by definition, are deep wounds. Not all stabbings are carried out with a very sharp instrument; many blunter implements, such as closed scissors or a poker, are sometimes used. Generally speaking, the blunter the instrument, the more ragged the entry hole is and the more bruised the surrounding skin becomes. When a sharp knife is used, the entry hole can be no more than a slit, as we have seen, particularly if the weapon is double-edged, like a dagger. A single-edged knife, such as a kitchen knife, will often leave a slit that tapers toward one end, as may be expected. However, the expected shapes of the holes or slits may be modified by the circumstances of the stabbing. If, for example, the victim twists, or the perpetrator turns the knife in the wound, the resulting entry hole will look very different. Therefore, it is not always easy to decide during the post-mortem what the width of the blade was. As regards the length of the blade, one is on firmer ground, for the depth of the cut (or the deepest cut, if there are several) will give a measure of the minimum length of the blade – minimum, because the blade may not have gone in right to the hilt. Similar reasoning applies to the estimation of the width of the blade; if wounds of various widths are present, then clearly the one with the narrowest width will give the surest indication of width. Again, one must interpret narrow wounds with care, since they may have been inflicted very shallowly, with only the blade's tip entering and cutting the skin. Such surface wounds are no guide to width of blade.

The way a stab is inflicted, or, more generally, the way an edged weapon is used, may give an indication as to whether the wound was self-inflicted or not. Cutting of the throat and stabbing through the heart are often a means of suicide, although this happens less frequently these days, due to the ease with which various poisonous substances are available. It is obvious that stabs in the back are unlikely to be suicidal, although I remember watching a whodunit film in which a man did stab himself in the back, not in a suicide attempt, but in order to mislead others into believing that he himself had been attacked. Having committed murder, his purpose was to deflect suspicion from himself by making it appear as though he, too, was the victim of an attack. I am glad to say that he was found out!

In cases in which the victim's throat is cut, there are several signs that point to the death as being suicidal or homicidal. Suicides never kill themselves with one clean cut of the throat. There are always several "experimental" cuts before the final, fatal cut or cuts are made. Dozens of cuts may be made experimentally, and some suicides have been known to give up the attempt after several trials, resorting to poison in the end.

In a suicide the cuts on the throat are shallow at first; and they are close together and more or less parallel to one another on that side of the throat opposite to the hand that is normally used for writing. The fatal wound or wounds are deeply cut downwards, then across the throat, then upwards again toward the other ear. These wounds can be very deep indeed, with knife-strokes cutting through to the spine; cases are known in which the head was almost completely severed.

Interestingly, people committing suicide in this manner often make it harder for themselves by tilting the head back, which causes the carotid arteries to slide back on either side of the spine. This means that most of the blood lost is venous, not arterial, which, in turn, means that the subject remains conscious for longer and that the agony is prolonged.

In homicidal throat-cutting the cuts are fewer and more cleanly incised; and the hesitant, experimental cuts are absent. Also, the cuts usually occur either higher or lower than the cuts made by suicides. Often, the victim's hands will be injured, as a result of the defensive action taken to fend off the attack.

If it seems odd that suicides can continue hacking at their throats for a long time without killing themselves instantly, it is worth making the point that many injuries usually regarded as being instantly fatal are, in fact, not always so. One such kind of injury is the stab to the heart; in one case a man stabbed in the heart was able to run to a window and jump out. In another case, a suicide who had more or less severed his head was still able to resist attempts to make him desist, to the extent that he was able to kick an ambulance man down the stairs!

An interesting case concerned with a heart stabbing began on July 3, 1949, in Klagenfurt, Austria. Squadron Sergeant-Major Slim Williams of the Royal Corps of Signals had asked a girl named Margaret to marry him. She was drunk at the time and accepted the proposal, although she realized later that she did not want to marry him. However, feeling bad about changing her mind, she had allowed the wedding to go ahead – in April of that year.

On July 3, they both went drinking at the Sergeant's Mess, where Margaret became drunk. On their way home they began to quarrel and she walked off on her own. Meeting an Austrian acquaintance, they went off to a Gasthof, where Margaret drank and danced. Soon, Slim arrived to find her, and he was furious. With him were two soldiers, who took her arms and held them behind her back and marched her to a truck and threw her in. She was held face-down, still with her arms held behind her back, but when the truck reached their home, she was taken in and pushed through the door. As soon as she was released, she turned round and kicked one of her captors, whereupon Slim closed the door and proceeded to beat her. She said he had never struck her before and she told him that she would sleep on her own on the sofa.

He began to berate her and she told him she would leave him, whereupon he began to abuse her verbally, but she turned away from him to go to the sofa. He took her by the shoulder and turned her round, slapping her across the face. It was then that she saw the knife lying on the table. She picked it up and told him she would knife him with it, if he tried to beat her again. He mocked her, saying he was not afraid of her little knife, and he came and started slapping her again. She struck out at him twice with the knife, once near the shoulder and a second time below his ribs. He said, "That's done it," or words to that effect, turned, walked across the room and sat down. She did not think anything was seriously wrong and went and sat on the sofa. Then she noticed that Slim's face had turned grey and that there was blood on the floor and on his trousers. She tried to stem the flow of blood and tried to warm his face, which had gone cold. He groaned, but did not speak. She went for help, but he died a little later.

This, in essence, was Margaret's story. A young medical officer, Lieutenant John McIver of the Royal Army Medical Corps, carried out the autopsy and concluded that the second stab wound – the fatal one – was inflicted after Williams sat down. This put a very different complexion on the whole business. There was a trail of blood from where Slim and Margaret had been standing to the chair on which he sat down. The distance was about seven or eight feet. The first wound had bled externally, but the second wound had not, although it had gone through the heart. There was blood on the chair and the floor behind it.

Was this one of those cases in which a man, stabbed through the heart, could walk across a room to sit down? The wound to the heart in this case was severe, causing a gaping hole through which blood would have escaped internally in great quantities. Moreover, the instantaneous collapse of blood pressure from such a wound would have stopped any bleeding from the shoulder almost immediately. Also, the fact that the heart wound had not bled externally must have been because the victim was seated or lying down.

All this was damning evidence against Margaret. But there was more. Margaret had said that Slim had removed his shirt and vest before she stabbed him. In fact, both garments bore a tear or slit that corresponded to the shoulder wound, but there was no similar tear in either garment to correspond to the heart wound. It seemed clear that Margaret had stabbed Slim when he was fully dressed; he then went to sit down, removed shirt and vest and was then stabbed again as he sat.

Margaret Williams was found guilty of murder and was sentenced to death, but the Home Secretary later granted her a reprieve.

GRAZES AND ABRASIONS

So far, we have considered what are called incised wounds, but there are, of course, many other kinds of injury. It is sometimes assumed that deep, complex injuries provide more clues to the forensic specialist, but, in fact, it is the simpler, non-fatal wounds that usually yield most information. Grazes of the skin, or abrasions, are considered to be the most informative of all wounds.

The reason that slight surface injuries often yield so much information is that they can show the marks of an attack, including a self-inflicted one, very clearly. For example, the marks left by the fingernails of a strangler on the neck of his victim may be mild and would certainly not have caused death, but can be recognized very easily. Finding such impressions could suggest that the examination of material from under the nails of a suspect is indicated, since blood or tissues from the victim may still be found there. Such material may be identifiable as belonging to the victim. Fingernail marks may sometimes be long and drawn, which may be caused by the victim's attempts at escape or at forcing the strangling hands away. We have already seen how bite-marks may be important clues in identification.

Bruises are another kind of injury; they are caused by the diffusion of blood from broken blood vessels into the surrounding tissues, where it clots. Bruises, or contusions as medics call them, are very often not accompanied by any injury or break of the surface skin, although this is not always the case. A stab from a blunt instrument may result in an incision and a bruise around the edges of the cut itself. Unfortunately, the size and shape of a bruise is not a reliable guide to its cause, since much depends on the force of the blow, the kind of tissues that received the blow, the age and medical condition of the victim and so on. However, it is worth disposing of one particular myth, namely, that bruises cannot be inflicted on a dead body. They certainly can be, since a bruise results from the seepage of blood from broken capillaries into the tissues. After death there is no heart pressure, so the blood seeps out to a lesser extent. Microscopic examination of the bruised tissues can reveal whether bruising took place before or after death. A bruise inflicted during life will contain a higher than usual count of white blood cells, which move to the injured site to start the healing process. Such a reaction does not take place if the bruise is inflicted after death.

Occasionally, unusual abrasions appear on the skin; and they can be extremely informative. In the case of Neville Heath, who was described as being sub-human, one of his crimes was identified on the basis of the marks of a diamond-weave whip on the body of a woman who was devoted to masochistic practices. Unfortunately for her, Heath not only obligingly whipped her, but he inflicted a number of terrible, Ripper-like wounds on her body, before strangling her to death. The whip-wounds had left the shape of the weave on her skin; when the whip was put against the body, the pattern fitted exactly.

Heath was one of those people whom no-one would ever think was a murderer. He was twenty years old when he was hanged; he was suave, handsome, fair-haired, fresh-faced, well-built and self-controlled. Just before he was led to the scaffold, he asked for a whisky. When it was brought to him, he said: "I think I'll make it a double".

The diamond-weave whip, together with other evidence, identified Heath as the killer. However, as so often in forensic science, things are not always what they seem, as the following case will make clear.

THE TELL-TALE HAND

Late in November, 1943, a woman named Rose Ada Robinson was found dead in her bedroom in Portsmouth. She was the licensee of a public house and was in the habit of taking the day's earnings up to her room in two bags every evening. Ill-advisedly, she was also in the habit of telling everyone that she did this. When the police entered her room they found the bags empty; they had contained £450. The time of death was estimated as being at about 2.00 am or 3.00 am. The window was broken, but the door was unbolted and unlocked, which clearly meant that the murderer came in through the window and left through the door.

Mrs Robinson was strangled to death. Bruises showed that she had struck her head against the window-sill, before being dragged to the floor and killed. Yet the marks on her neck were rather unusual. There was one large bruise over the voice-box and three smaller ones to the left as one looked down at the neck (i.e. on Mrs Robinson's right). Oddly, there were no fingernail impressions where one would have expected them to be; but there were scratches that had clearly been caused by Mrs Robinson herself as she struggled to free herself. Odd, too, was the small span of the hand that strangled her – four inches.

What could it mean? Suspects were questioned and cleared. Some time later, the police arrested a man who was trying to sell some stolen shoes. His name, he said, was Harold Loughans. No doubt to the officers' surprise, he began to confess to other crimes. Weeping, he said that he had killed

Mrs Robinson a month earlier. Interestingly, the tips of all the fingers of his right hand, except the thumb, were missing. This fitted in with the marks, suggesting a small hand, on Mrs Robinson's neck.

When the case came to trial, Loughans denied his confessions and said that they had been made up by the police. An important part of the evidence concerned the question of whether Loughans had enough strength in his deformed right hand to strangle someone to death. The prosecution's pathologist, Professor Keith Simpson, believed that he had, but the defence pathologist, the great Sir Bernard Spilsbury, did not. Spilsbury had asked Loughans to grip his hand as hard as possible and, on the basis of this, concluded that his hand was too weak to strangle anyone. Both pathologists gave evidence, but, such was the awe in which Spilsbury was held, the jury found Loughans not guilty.

When Loughans left the court-room, he was immediately arrested by the police for another crime, for which he was convicted. In prison he became known as "Handy", because of the dexterity of his right hand. Years later, the newspaper, *The People*, published an article, saying that Loughans was, in all probability, guilty of Mrs Robinson's murder. Loughans took the newspaper to court for libel, an action he lost. A few months later, Loughans knew he was dying of cancer. He walked into the offices of The People and made a written confession: "I want to say I done that job. I did kill the woman in the public house in Portsmouth". He wrote the confession with his right hand.

DEATH BY ASPHYXIATION

Death by strangling is due to asphyxia – the inability to breathe because of obstruction of the respiratory apparatus. The biochemical changes that occur include a decrease of oxygen in the blood and tissues, together with an increase of carbon dioxide, and the development of a suffused, purplish colour – cyanosis – in the skin. Blood-staining may appear around the nose and mouth. Petechiae, or spots of haemorrhage, appear, due to the bursting of small capillary blood vessels. These petechial haemorrhages are particularly good signs of asphyxia. However, sometimes the inhibition of the vagus nerve (because of the pressure applied to it) during strangulation may cause death suddenly and quickly, before the usual symptoms of asphyxia appear. In such cases the natural tendency to struggle to breathe does not happen.

People being strangled often vomit, increasing the probability of asphyxia, since the vomited matter may clog up the air passages further.

However, in cases in which vomiting has occurred, it is important to consider the possibility that the vomiting may not have been caused by strangulation, but that the act of vomiting itself caused the asphyxiation of the victim. In other words, vomiting could be the cause, rather than the consequence.

Clearly, asphyxia can be caused in a number of ways, such as by choking on food, suffocation (in which the victim's mouth and nose are covered), or by what is known as overlaying, as when a mother rolls over in bed and accidentally comes to lie on top of her baby. Many such cases are clearly accidental, but overlaying is sometimes treated as a criminal matter, even if it is accidental. This is the case when someone had been drinking and then went into bed where a baby is sleeping. Overlaying in such a case is a criminal offence.

A particularly common form of death caused by asphyxia is, of course, drowning. When someone drowns, the air passages, as well as the stomach are filled with water. However, due to the pressure of water in the lungs, the blood vessels tend not to burst and rarely do petechiae appear. Also, people who drown in natural bodies of water will often have large numbers of diatoms – microscopic algae with silicon shells – in their blood, bone marrow and brain. Debris from the water, such as soil or small water organisms, will be found in the stomach. Drowned persons will show froth emanating from the mouth and nose.

Not all water deaths are, strictly speaking, drownings. Vagal inhibition, caused by a sudden change in temperature, such as falling into icy water, may cause death before drowning takes place. It is often said that those who died when the Titanic sank were drowned; in fact, they died of the cold.

Our bodies have a particular osmotic balance, which it is harmful to alter to any great extent. The inhalation and swallowing of water during drowning, therefore, results in great disturbance to this osmotic state, with severely damaging consequences. Interestingly, drowning in fresh water is usually very rapid, since the blood is diluted to such an extent that water will enter the blood cells, causing them to burst. In sea water, drowning is much slower, since the increase in osmotic pressure in the blood will cause water to flow out of the blood cell into the plasma, a state with which the body can cope much better.

Drowned people found unclothed or wearing a swimming costume are likely to have drowned by accident, although this is not necessarily always the case. People found drowned while fully-clothed are more likely to have drowned by intent, usually murderous intent. Suicides tend to remove some of their clothes before taking the plunge.

People can sometimes drown in very shallow water, although foul play may not necessarily be the cause. Injured or sick people may collapse and die in situations that look suspicious, and it is not unknown for such people to fall into, say, the edge of a lake or river and drown in a foot or two of water.

Whenever I have lectured about forensic science to students or general audiences, I have noticed that there is a general assumption that the more scientifically complex a forensic investigation is, the more impressive is the evidential value of the results. In fact, the opposite is often true; a simple, straightforward clue, correctly interpreted, may shed a good deal more light on a case than can a clue that requires investigation using sophisticated and expensive equipment. A historic drowning incident is a case in point.

On the afternoon of June 13, 1886, Ludwig II, King of Bavaria, left his castle, Schloss Berg, and went for a walk with his physician, Dr Gudden, by the shores of Lake Starnberg. They were expected back at the castle by eight o'clock in the evening but, when that hour arrived, they had still not returned.

An hour later, a search was started. Before long, the searchers found ominous signs that a dreadful thing had happened, for on the path that went round the lake the king's umbrella and his jacket and overcoat were found. It had been raining heavily and the clothes were drenched. The jacket and overcoat must have been removed in a hurry, since they were both turned inside out, with the sleeves of the jacket still inside those of the overcoat.

Close to the spot where the clothes were found were several bushes and trees. Some branches were broken, suggesting that someone had moved quickly, or been pushed, through the vegetation. Beyond the bushes was the lake. Soon, the search party found what it was looking for; the dead body of the king was found floating face-downwards, with his arms stretched out toward the shore. The water was less than four feet deep at that spot. Not far from the king, Dr Gudden's corpse was found floating; although he too was face-down, he was in an odd half-kneeling, half-sitting position.

What had happened? Was it accident, suicide or murder? The first possibility can be ruled out; strong men who are also strong swimmers do not drown in four feet of water; and the king was an exceptionally strong swimmer. The full story of King Ludwig is too complex to recount here; suffice to say that he was surrounded by political intrigue and that many people wanted to see him deposed. The king himself was a very unstable character. In short, both murder and suicide were possible under the circumstances.

The accounts of who saw what and when are all so confusing and unclear that it is difficult to arrive at the truth. Some say that there was a bullet wound in the king's head; others say not. Accounts differ about what happened immediately after the discovery of the bodies. Some have said that the king might have died of a heart attack, but this does not explain Gudden's death. In any case, there is no evidence of a heart attack. The clothes were burnt in secret, at dead of night. There does not even appear to be any evidence that the manner of death was drowning; none of the accounts mention the usual signs of drowning. The king's face was said by some to have worn an angry, tyrannical expression; others made no comment on this. One witness said that something was later "done" to the king's face, making it look like a mask. Again, others have said nothing about this. Some have asserted that the king murdered the doctor, then killed himself; others have ridiculed the idea.

So, is it at all possible to find out what happened that day? Perhaps not in exact detail, but there is one piece of evidence that is quite undisputed and which has been overlooked as a clue in all the debates about the way King Ludwig met his death.

It is simply this: his bowler hat was found floating on the water. So, too, was Dr Gudden's silk top hat. Now it is an interesting fact that people who commit suicide invariably remove their hats before they throw themselves into the water. Perhaps there are exceptions to this rule, but I have not come across any and they must be very rare indeed. Also, deep water is chosen for suicidal purposes, as well as a high jump. Choosing less than four feet of water in which to drown oneself is absurd.

But what of the jacket and overcoat, how can their removal be explained? We have seen that they were removed in a hurry and turned inside out; this suggests that they might have been removed to escape the clutches of someone gripping the king by his clothes. When this happens, the clothes tend to turn inside out and are left in the assailant's hands. The king may then have reached the lake, with his bowler still on his head, was then pursued into the water by his attacker and drowned or shot. It is the only scenario that covers all the undisputed facts. Accident can be ruled out; suicide is extremely unlikely – what is left is murder.

Asphyxial deaths are not always easy to recognize, and even the greatest forensic specialists can be misled by signs and circumstances that seem to indicate asphyxia by foul means. The story of Sidney Harry Fox is a case in point.

Fox was a forger, a blackmailer and a thief. He was also a homosexual, a matter that was frowned upon much more in 1929 than it is today. (This point is important, in view of what happened later.) His mother, an old woman who was far from well, was his accomplice in crime and travelled around the country with him – they were never in one place for long and spent most of their time in one hotel or another, hardly ever paying their bills when they left.

On October 22, Fox left the Hotel Metropole in Margate and went to London, where he took out two insurance policies on his mother's life, then returned to Margate. At about 11.30 pm, Fox ran downstairs, shouting that a fire had broken out in his mother's room. Someone ran up the stairs with him and managed to pull Mrs Fox out of the smoke-filled room. They tried artificial respiration, but, in spite of their efforts, Mrs Fox died.

Two doctors arrived at the scene, one after the other. Dr Austin arrived first and, after examining the body, pronounced that the deceased had died as a result of shock and suffocation and signed a death certificate to that effect. When the second doctor – Dr Nichol – arrived, he did not question Dr Austin's conclusion. The inquest resulted in a verdict of accidental death.

Matters, however, did not end there. Fox claimed the insurance, but one of the insurance companies was suspicious and approached Scotland Yard. The body of Mrs Fox was exhumed and a post-mortem was conducted by Sir Bernard Spilsbury. His conclusion was that the deceased had been strangled to death. Fox, of course, was charged with murder and his defence lawyers engaged the services of that other great Titan of forensic pathology, Sir Sydney Smith.

There was little or no sign of manual strangulation on the body. No petechiae, no fingernail marks, no injury of any kind to the neck, the mouth or the nose. In particular, the delicate hyoid bone was not broken, although it is very easily broken during strangulation, especially in elderly people, in whom it is very brittle and fragile. In fact, Spilsbury himself could not help breaking Mrs Fox's hyoid bone during the post-mortem, it was so brittle. Although, as we have seen, the classical signs of asphyxia are not always present in cases of strangulation, this does not, of course, mean that their absence indicates that asphyxia had taken place – that would be perverse logic indeed.

Spilsbury's conclusion that the cause of death was asphyxia due to manual strangulation was based upon the facts that the epiglottis (a fold of cartilage at the base of the tongue) was haemorrhaged and that there were three bruises – one each on the tongue, the larynx (or voice-box) and the thyroid gland. Spilsbury maintained that these bruises could, somehow, have been inflicted by manual strangulation despite the fact that there were no bruises or abrasions on the skin of the neck. In particular, he thought that the bruise to the larynx was significant.

According to Spilsbury's post-mortem report, Mrs Fox's heart was at an advanced stage of deterioration and she had cirrhosis of the liver. It was clear that a shock could easily have killed a woman in that state of illness. Nevertheless, the great pathologist stuck to his guns.

It was arranged for Sir Sydney Smith to examine the preserved larynx in Spilsbury's laboratory. When the specimen was produced, Smith could see no bruise, nor could his associate Dr R.M. Brontë. Spilsbury said that the bruise had been there, but had now "faded". This is quite impossible, since the discoloured blood that forms the bruise will not disappear, unless putrefaction had progressed, which was not the case here, as the larynx had been preserved in formalin after the post-mortem.

The other bruises were also examined by Smith; the one on the epiglottis was the size of a pin-head and would have been present in most deaths of natural causes. The bruise to the tongue was certainly of a kind that could have been produced by manual strangulation, but could equally easily have been caused by someone biting their tongue – especially if they had badly fitting dentures, like the ones Mrs Fox wore. Finally, the bruise on the thyroid consisted of "a few stray red blood corpuscles", in the words of Sir Sydney Smith and would be found on any normal thyroid.

Yet Sir Bernard Spilsbury did not retract his evidence. It is said that he never admitted a mistake and never changed his opinion once he had given it. As we have seen, he had enormous "presence" and the jury believed him. Fox was found guilty of murdering his mother and was hanged, although he was almost certainly innocent.

NO-ONE IS INFALLIBLE

At this point it is worth making an aside that is relevant to the practice of forensic science. The two cases involving Sir Bernard Spilsbury that we have looked at so far show him to be both mistaken and inflexible, yet he was one of the greatest forensic practitioners of the twentieth century, as we shall see later. Nevertheless, greatness carries with it the risk that the man in question may acquire such a degree of confidence and belief in himself that it becomes well-nigh impossible for him to admit to making mistakes. I have seen this in a number of forensic witnesses.

Indeed, I will confess to recognizing this in myself, not, I hasten to add, because of any delusions of greatness, but simply because I was for a long time the only practitioner in my field of forensic specialism in Britain. For years no-one could be brought into court to question my opinions, with the consequence that my testimony always went unchallenged by another specialist, although I was often stiffly cross-examined by barristers. This is an unqualifiedly bad thing, for the simple reason that no-one is infallible; and I certainly am not. I wrote to the Home Office several times, explaining my concerns in this matter and urging them to train another practitioner in this field, so that there would be at least two opinions available to the courts. Unfortunately, the Home Office did not see the need for this course of action at that time. Since then, with the increasing awareness of the usefulness of forensic entomology, several new practitioners have emerged.

There is another salutary lesson to be learnt from the Fox case, and it is this: forensic scientists are human. They have their prejudices and their fears. Fear of being mistaken is a very real thing among scientists of all kinds, and prejudices against those of whom we disapprove is by no means a rare thing. Fox was a homosexual, which, no doubt, biased people against him. He was also a thief, a liar and a swindler, but these things should not affect one's opinions on his guilt or innocence of any particular crime, only the relevant evidence should be considered.

Although these comments may seem too obvious to mention, or even seem too self-righteous and smug, they do deserve airing. If one believes that justice should be given only to those people of whom one approves, then one does not believe in justice. There can be no other logically or morally coherent opinion on the subject – you must take it or leave it.

DEATH BY HANGING

Hanging is a particular manner of death, of which asphyxia is only part. It is important to remember that the word "hanging" is used to describe two somewhat different manners of death. A judicial hanging, or execution, involves a long drop that breaks the neck. This seldom happens in non-judicial hangings, most of which by far are suicides. A suicide hanging causes death in three ways: first, the pressure on the jugular veins and carotid arteries results in a lack of oxygen reaching the brain; secondly, the pressure on the vagus nerve causes breathing inhibition; and, finally, asphyxia results, because the breathing passageways are obstructed by the tongue and glottis, which are pushed into the pharynx.

Victims of hanging show much the same post-mortem signs as do victims of strangulation, notably cyanosis and petechiae. In addition, the tongue is often found protruding through the mouth and the eyes may bulge out, and ligature marks will, of course, be present.

Although very few hangings indeed are homicidal, some occur by accident. I know a case of a 12-year-old boy who hanged himself accidentally while playing with a rope on the stairs, tying one end of the rope round the bannister and the other end round his neck. He slipped, the noose tightening around his neck, probably inducing him to lose consciousness very quickly before dying. However, most accidental hangings take place when young men indulge in certain masochistic practices. It seems that these practices and their consequent deaths are much commoner among men than women.

Death by hanging is not always caused by all three factors listed above. When, on March 7, 1975, the body of Lesley Whittle, the victim of Donald Neilson, better known as the Black Panther, was found hanging sixty feet below ground level in a concrete shaft, it was first concluded that she had died of asphyxia. Later, it was found that vagal inhibition was the true cause, her heart stopping through sheer terror.

BLUNT INSTRUMENTS

We come now to the use of the blunt instrument as a murder weapon. Murders committed this way are usually unpremeditated killings and may not be murders in the strict sense at all, but such things as crimes of passion or assaults committed under provocation. However, premeditated murders are committed with blunt instruments, and can be among the most horrific of all kinds of murder.

To begin with, an assault with a blunt weapon is usually made at the head, not at other parts of the body. Death is often caused by brain injury and the scene of the crime is almost always very bloody, since head injuries bleed copiously.

The "effectiveness" of a blunt instrument depends on two attributes: its weight and the speed with which it is wielded. The resulting force of the blow – its kinetic energy – can be expressed as follows:

$$E = \frac{m \times v^2}{2}$$

Where E = kinetic energy, m = mass and v = velocity.

It is clear from this equation that the speed with which the weapon is travelling at the time contact is made has the greater effect, since its value is squared. In other words, a strong blow with a piece of wood will cause more damage than a feeble blow with a sledgehammer, speaking in very general terms. Among the commonest blunt instruments are hammers, pokers and other household implements.

Although blunt instruments may be seen as being the least "professional" of murder weapons, to my mind they are among the most unpleasant. This is not only because of the large amount of blood that is splashed about the scene, but because several well-aimed, powerful blows are required to kill the victim. A stab from a knife or a bullet from a rifle will, usually, kill instantly, whereas the wielder of the blunt instrument displays a kind of brutal tenacity that is particularly dreadful.

Once the skull is cracked or smashed by a blow, damage to the brain and its membranes will occur. In fact, damage to the membranes can occur even if the skull is not fractured by the blow. Epidural haemorrhaging (bleeding outside the dura mater, the tough membrane that lines the skull) can occur due to the tearing of the blood vessels supplying the brain. This type of injury is most commonly seen in teenagers, since as well as being caused by violence, it can also be the result of a blow sustained in a traffic accident or in a fall especially as a sporting injury, but it can happen to anyone. For example, the actress Natasha Richardson sustained a fatal blow to the head after falling during a beginner skiing lesson. The blood accumulates between the dura and the bone of the skull. Often, no injury, apart from bruising of the skin, is apparent, whereas, in fact, the blood is accumulating in the space between membrane and skull, putting pressure on the brain. Death may follow a few hours later; post-mortem examination will show a blood clot at the site of bleeding.

Subdural haemorrhages (bleeding between the dura and the brain itself) often result in a blood cyst. As blood leaks very slowly in this type of injury, it is rarely fatal and may go unnoticed for years. What are known as subarachnoid haemorrhages (bleeding below the arachnoid membrane of the brain) are as often caused by natural causes as by blows to the head and are often associated with aneurysms (dilations of the arteries, due to the weakness of the arterial wall). Sometimes a relatively light blow may cause haemorrhaging. In one case a father, fighting with his grown son, was horrified to see the young man collapse and die after he had struck him a few light blows. The post-mortem showed that the son had an aneurysm and the father was discharged. Subarachnoid bleeding in the absence of an aneurysm gives grounds for suspicion of foul play.

Blows to the head, even with the fist, can cause damage to the brain as a result of the movement of the brain within the cranium. A blow in itself may not cause much damage, but if the head twists round at some speed as a result of the blow, the brain will be chafed against the inside of the skull, since the latter will be travelling faster than the former. In this context, I have often been puzzled by the fist-fights that are commonly shown on film, typically in cowboy westerns. The blows inflicted by these people on one another would, in reality, cause severe external and internal damage, although in the films they hardly ever sustain even a minor bruise! This would be merely amusing, were it not for the possibility that youngsters, emulating their film heroes, may injure themselves quite seriously.

Another type of head injury, known as contre-coup, results when the moving head strikes a stationary object, such as hitting the ground after a fall. Here the injury takes place at a spot opposite that at which the head was struck. If a man falls backwards, striking the back of his head on the floor, injury to the front of the brain may occur. This is because the skull is travelling much faster than the brain, which will receive chafing injuries, as well as "piling up" injuries to the front of the brain.

It is quite remarkable how injuries to the brain can sometimes fail to result in death. One case on record involved a man who drove a four-inch nail into his brain through his forehead in a suicide attempt. The nail was removed and the man survived!

Injuries inflicted after death are usually easily recognizable as such, since there is no cellular reaction, as we have seen in the case of post-mortem bruises. Injuries inflicted by blunt weapons are no exception, as was shown very clearly when the battlefield of the Little Bighorn River in Montana, was excavated by archaeologists. This battle – in which General Custer, at his last stand, was killed, together with all his men – was followed by some horrific cruelties. The Sioux and Cheyenne Indians were apt to mutilate the dead and the nearly dead after a victory. Black Elk, the warrior chief, described how his warriors used clubs and hatchets to mutilate the dead and finish off the injured survivors. Interestingly, the anthropologists working on the archaeological dig were able to confirm his story and showed how the bones of so many of the bodies had been broken while the bone was still "green" or fresh since the fracture edges were sharp and clean (whereas "dry" bones would show jagged and torn edges at breaks). Such injuries, inflicted at or about the time of death, have been termed perimortem injuries.

It is fortunate that it is possible to distinguish between injuries inflicted after death and those inflicted before death, since much can turn on such matters. Injuries are sometimes inflicted after death in order to suggest that death was due to suicide, not murder. A case in point concerns the death of Sir Edmund Berry Godfrey, who was found impaled on his sword in 1678. Initially, it was believed that he had killed himself, until the post-mortem examination revealed that he had been strangled to death, the sword having been thrust through him after death.

It has often struck me that murders with blunt instruments are not only among the most brutal of crimes, but also the most senseless. The story of San Dwe is a case in point.

San Dwe was a Burmese elephant keeper who worked at London Zoo during the 1920s. His special responsibility was a white elephant, but it soon became clear that the English climate did not suit the animal, which was sent back to Burma. San Dwe moped and fretted, missing his white elephant dreadfully. However, this was not his only grievance. Another keeper, an Indian named Sayed Ali, who, like San's white elephant, could not cope with the English winter, was allowed by his employers to return to India during the winter and come back during the summer, since he was a very good elephant keeper and the zoo authorities valued him highly.

When Sayed Ali was in India, San Dwe stood in for him and benefited from being the recipient of the magnificent weekly sum of thirty shillings in tips when he took children for rides on the elephants. Of course, when Sayed Ali returned in the summer, he took over once again and San Dwe lost his lucrative business. On August 4, 1928, two policemen heard groans coming from the direction of the Tapir House, which lies close to the Outer Circle of Regent's Park. One of them climbed over the railings and found San Dwe lying on the floor, with a wound in his foot. He was raving and shouting, but the police managed to extract the story from him. Four men, he said, broke into the house and killed Sayed Ali with a pickaxe and assaulted San Dwe himself, wounding him in the foot.

The police found Sayed Ali's body, as well as a pickaxe and a sledgehammer, both bloodstained. His head had been fearfully battered and his room had been ransacked. There was a wallet, containing £36 in cash, lying on the bed. When questioned by the police, San Dwe said that he desperately wanted to return to his country. It was clear that he had killed Sayed Ali, with the aim of robbing him and using the money to pay for his passage to Burma. He lost his nerve after the murder and left some of the money behind.

At the trial, no-one believed in the existence of the four men. San Dwe was found guilty, but the sentence was commuted to penal servitude for life. In 1932, he was released and repatriated to Burma. The saddest thing about the whole affair is that, if he had only told the zoo authorities of his wish to return to his country, they would have released him and paid his way. San Dwe was, in effect, a child, as far as his mental capability was concerned, and it was this that saved him from the noose.

The crime of William Podmore, which took place shortly after San Dwe's, was considerably less excusable. Vivian Messiter's body was discovered in a garage in Southampton in January, 1929. At first, it was thought that he had been shot; then it was believed he had died of a haemorrhage. In fact, he had been battered to death with a hammer.

It is worth pointing out at this juncture that, contrary to popular belief, most uniformed police officers never see a case of murder and their initial description of the scene may leave a great deal to be desired. Despite the fact that Messiter's head had been severely battered, the above conclusions about manner of death were suggested.

The full story of this case is a very involved one, but the essential point is that Messiter, who was an agent of the Wolf's Head Oil Company, discovered that Podmore was embezzling company funds. When Messiter confronted Podmore with this, there followed an argument, which ended in murder.

The interesting findings of Sir Bernard Spilsbury, who examined the body, were as follows: there were several blows to the head with a blunt instrument like a hammer; one of the injuries was inflicted on the forehead, just above the eye, by the pointed, back end of the hammer, the others by the flatter, heavier end; the hammer had to have been of the type that had a pointed end, not the usual kind of hammer with two curved prongs for the extraction of nails; one injury was to the back of the head, suggesting that the victim had been either bending over or was struck from behind.

It was the hole-like puncture above the eye that gave rise to the idea that Messiter had been shot. Whatever suggested a haemorrhage remains a mystery, but, as we shall see later, such bizarre conclusions are not unknown.

A blood-stained hammer, of the type Spilsbury described, was found in the garage not far from the spot at which Messiter fell. But the most damning of Spilsbury's conclusions was that two of the injuries were inflicted when the man was down, with his head resting on a hard surface. This weighed heavily against Podmore in court, suggesting as it did a callous vindictiveness. Another clue was the final nail in Podmore's coffin. Two hairs were found by Spilsbury adhering to the hammer. These were compared with the victim's eyebrows and were found to be identical. Podmore was found guilty and was hanged.

Even greater brutality than Podmore's can be expressed by those who wield the blunt instrument. An example is one of the most inhuman crimes imaginable – the case of three-year-old Marion Ward.

In August, 1949, Marion's body was found among the ruins of a bombed house; her head had been beaten to a pulp. The pathologist, Francis Camps, concluded that there had been nine separate blows to the head. That anyone should do such a thing to a small child was appalling enough, but suspicion soon fell on a couple, Mr and Mrs Tierney, who were themselves the parents of a six-year-old girl, Stephanie, and a small baby boy, Jamie. Stephanie often played with Marion. James and Nora Tierney were well known in the neighbourhood as an unhappy couple who frequently had rows. It was also known that Mrs Tierney had once attacked her husband with a knife, although she denied this and maintained that it was he who was violent.

Interestingly, and rather bizarrely, the husband and wife accused one another of the murder when questioned by the police, who were by now certain that one of them was the murderer. A thorough examination of the scene revealed the footprints of Mrs Tierney; but there was no evidence that Mr Tierney had been there. It seemed clear that Nora Tierney was the guilty party, but the most damning evidence against her came from the most unexpected source.

There was an eyewitness to the crime, none other than Stephanie, the Tierneys' six-year-old daughter. She told the police that her mother had tried to gag Marion with a handkerchief, but the child struggled to pull it off and called for her mother. "My Mummy took a hammer out of a brown handbag," Stephanie told the police. "Then Mummy took the handkerchief off Marion's mouth. Marion fell down and Mummy kept hitting her on the head. Marion's eyes were still open and Mummy said: 'She is not dead yet. She is hard to kill.'"

A crime, horrific enough, was witnessed by the child of the woman who committed it. Such horrors hardly need any further comment, although we will return to this story in the final chapter of this book. There was no apparent motive for the crime; and Nora Tierney was found guilty, but insane, and was sent to Broadmoor.

DEATH BY SHOOTING

Firearm killings are commoner in the United States than in Britain and western Europe, although they are increasing on this side of the Atlantic. Essentially, there are two types of firearm: smooth-bored ones and rifled ones.

Smooth-bored weapons, better known as shotguns, have barrels that are smooth on the inside; and they fire shot pellets rather than bullets. The barrel is usually narrower toward the muzzle end, which helps to keep the pellets together.

Rifled weapons are not smooth on the inside of the barrel, which is spirally scored or "rifled". This ensures that the bullet rotates on its axis as it travels through the barrel and outside it; this helps to stabilize the bullet in flight and prevent it from wobbling, enabling it to travel on a straight course for up to one thousand metres. The inside of the barrel will, therefore, be made up of raised spiral bands (the lands) and indented spiral bands (the grooves). Rifles, revolvers and pistols are all rifled weapons and they fire bullets, which are typically solid lead hardened with tin, or have a lead core coated with steel, which is itself coated with cupro-nickel. A number of other kinds of bullet exist, but they need not concern us further here. The calibre of a gun is its diameter from land to land (not groove to groove) in inches. The types .303, .45 and so on refer to this diameter. It is clear that the calibre should not be measured from groove to groove, since this will give a diameter greater than that of the bullet, which fits between the lands, not the grooves.

One of the standard acts so often seen in whodunit films is when the police inspector finds a handgun at the scene and picks it up gingerly by putting a pencil inside the barrel and lifting the weapon. This is meant to show that the inspector is trying not destroy evidence by holding the gun with his hands, in case he smudges any fingerprints on it. In fact, picking up a gun in this way is the very last thing one should do, because anything inserted into the barrel will almost certainly leave its mark on the "lands" of the rifling. This is important, since bullets fired from a particular gun will show markings corresponding to the rifling of the weapon from which it was fired. No two guns being identical, even those made by the same manufacturer in succession, it is possible to compare a bullet discovered at the scene or inside a victim, with a bullet experimentally fired from a suspect gun. This can reveal whether a bullet was or was not fired from that gun. For this reason, guns must be handled carefully – by someone wearing gloves – and placed in a bag for subsequent examination.

A particularly interesting case of bullet identification took place when President John F. Kennedy was assassinated. From then on many people believed that the crime was the result of a conspiracy and that the assassin, Lee Harvey Oswald, could not have acted alone. Others ridiculed the notion. Over the years something of the order of ten thousand books on Kennedy's assassination were published, some arguing the case for conspiracy, others arguing the case against it. The usual accusations and counter-accusations were made: "cover-up" on the one hand, "conspiracy theorists" on the other. What does the forensic evidence have to say about this?

The Warren Commission examined all the evidence, scientific and otherwise, and concluded that there had been no conspiracy.

Of course, there is always a small group of paranoid people who will always shout "conspiracy" on every occasion. In the case of President Kennedy, it must be said that there is a prima facie case for conspiracy. The idea is not an absurd one. Kennedy had many enemies at every level in society, including powerful political circles. The murder of Oswald himself has not been adequately explained. Eyewitness and other evidence also supported the idea of conspiracy. Some people who expressed their belief in the conspiracy theory did not survive for very long, although this line of "evidence" has been greatly exaggerated.

None of this means that Kennedy was murdered, but it shows that it is a valid suspicion. What I find particularly interesting is that the forensic evidence from the case has been held to reveal that there is no rational basis for supposing there to have been a conspiracy. The Warren Commission held that Oswald acted alone. I have reviewed the forensic evidence and find that I cannot agree with that conclusion.

This is not the place to go into a detailed examination of this evidence. The relevant facts are these: Oswald was said to have fired three shots, one of which totally missed the car; one struck Kennedy in the back, emerging from his throat to strike Governor Connally, who was sitting in front of him, in the back; the bullet left through Connally's chest striking first his right wrist, then his left thigh. The last, fatal, bullet struck the president in the head.

The recovered fragments of these bullets were examined for the trace elements of silver and antimony. The amounts of these elements found in a bullet vary in such a way that their proportions in the fragments can suggest the number of bullets from which they came. The results of the study of the fragments were held by the Warren Commission to show that only two bullets struck the occupants of the car. This is puzzling, since it is clear that the trace element evidence shows that more than two bullets were fired.

The results for traces of silver, in parts per million, were as follows (the ranges of the results are given in brackets):

1. Bullet from Connally's stretcher 8.8 ± 0.5 (8.3 – 9.3)
2. Fragments from Connally's wrist 9.8 ± 0.5 (9.3 – 10.3)
3. One large fragment from the car 8.1 ± 0.6 (7.5 – 8.7)
4. Fragment from Kennedy's brain 7.9 ± 0.3 7.6 – 8.2)
5. Small fragments from the car 8.2 ± 0.4 (7.8 – 8.6)

The range of trace silver in (1) and (2) just meet, which is inconclusive, since they could represent either one or two bullets. The ranges for the remaining three specimens overlap considerably and suggest that the fragments came from one bullet.

The results of the antimony analysis showed the following results:

1. Bullet from Connally's stretcher $833 + 9$ (824 – 842)
2. Fragments from Connally's wrist $797 + 7$ (790 – 804)
3. One large fragment from the car $602 + 4$ (598 – 606)
4. Fragment from Kennedy's brain $621 + 4$ (617 – 625)
5. Small fragments from the car $642 + 6$ (636 – 648)

None of the ranges of trace antimony in the bullets or fragments overlap, yet these results were used to demonstrate that only two bullets were fired, which is an impossible conclusion. Of particular interest is the fact that the two Connally bullets show very different ranges. What all this means is another matter. It certainly does not prove conspiracy. But this is not the point. The point is that, in spite of the fact that some of the evidence suggests that the accepted version of events is incorrect, that very evidence was used to "demonstrate" the opposite. Such things do happen in forensic science.

Despite stringent controls on gun ownership and the banning of most handguns in the UK, gun-related offences do occur. Many of the guns involved are illegal and tracing their route of supply is as important as solving any crime those guns have been linked to. As with much else in Forensic Science, the advent of computers and computerised imaging technology has had an enormous impact. In 2003, the UK's Forensic Science Service developed a National Firearms Forensic Intelligence Database, which stored details of the make, model, calibre, and any modifications or conversions, of the weapons used in incidents nationwide, together with a description of the type of crime committed. It could also store bullet and cartridge case characteristics in a similar way to that used for fingerprint files and used an automated system (the Integrated Ballistics Identification System or IBIS) to search for matches to a particular bullet, cartridge case or firearm. In 2008, the National Ballistics Intelligence Service (NABIS), was formed to provide a fully national database of all recovered firearm and ballistic material coming into police possession anywhere in the United Kingdom, together with tactical intelligence relating to people, objects, locations and firearm-related events nationwide. NABIS also runs four regional centres for test-firing, analysing and linking weapons and other materials to incidents across the UK. The aim is to be able to quickly provide information to investigating officers, which can be especially useful in cases where a weapon has been used in other crimes in different locations. As for fingerprint information, the automated database search can suggest matches between a bullet and a particular gun, or between bullets found at different scenes, but a definitive match that can be used as evidence in court must still be made by a ballistics expert and this is one service that NABIS deliberately does not provide in order to ensure that the final match will be made by an independent forensic expert. Nevertheless, once an item is on the NABIS database, there it stays, and from there it can be tracked through all stages of an investigation, whether in independent hands or not, until all proceedings are completed and it is eventually destroyed.

Nowadays, all legal firearms' manufacturers stamp a serial number onto every gun, and this can be very useful in helping to trace the history of that weapon. In cases where the original stamped number has been obliterated by rubbing or the rifling of the weapon, it is often possible to restore the number by the use of etching agents. This is possible because the die that stamped the number will have given rise to stresses in the metal beneath the indentation it caused. These stressed areas will dissolve much faster than the surrounding unstressed areas, showing up the number once again.Much has been written about the wounds inflicted by bullets.

Generally speaking, most textbooks say that the exit wound of a bullet is much larger and more ragged than the entry wound. Other books – a small minority – assert, equally authoritatively, that entry wounds are larger and more ragged than exit ones. Who is right?

Both and neither. Much depends on the circumstances of the case and the nature of both the bullet and the part of the body it went through. In principle, when a bullet enters the body and emerges from it, the entry wound is larger than the exit one, *if it is a normal bullet and if it is moving through flesh*. In fact, this rarely happens, since most bullets hit bone at some point during their "journey". When this happens, the exit wound is larger and more ragged than the entry wound.

An entry wound is often easily recognisable as such by the discoloration around it. Bullets are not the only things that emerge from the muzzles of guns; soot and primer residues are produced and, in contact shootings in which the muzzle is held against the body, the skin around the entry hole is blackened and may be pinkish in colour due to the emission of carbon monoxide. The hole itself will have a splintered or star-shaped appearance, since, despite the rifling, a bullet that has just emerged from the barrel will show some degree of wobbling, or tail wag, before it stabilizes. This tail wag tears the skin and produces the ragged-edge effect.

A shot fired from about a yard away will also show these characteristics, but to a much lesser extent, the hole being smaller and the "tattooing" of the skin much slighter. From a greater distance there will be little or no blackening and the exit hole will be roughly the same size as the entry wound. However, the margin of the wound will be soiled; this is because the bullet, spinning at a rate of two thousand and three thousand revolutions a second, will wipe off any material coating it on to the skin.

In cases in which the bullet strikes bone, the flattening of the bullet and the bone material it carries with it through the body will result in an exit wound that is much larger and more messy than the entrance wound.

The angle at which the bullet entered the body can sometimes be important evidentially. It is possible to gain some idea of the kind of angle involved by examining the blackening of the skin around the entry wound. For example, if the gun is held at an upward angle, the blackening will be more intense on the skin above the wound.

In the Libyan Embassy siege in London, during which a policewoman standing in the square outside was killed by a bullet, it was important to determine the direction from which the bullet came. The post-mortem findings revealed that the bullet entered the chest at an angle of between

60° and 70° to the horizontal. Examination of the pavement revealed bullet impact marks that further supported the autopsy conclusion about the angle of the shot. The general conclusions showed that the bullet was fired from the first floor of the embassy. Later, an examination of one of the first floor rooms facing the square revealed a fired cartridge case of the right calibre, it having been missed by the perpetrators when they attempted to clear up the room.

The superficial appearance of shotgun wounds is, in certain respects, the exact opposite to that of bullet wounds. Contact or near-contact shots will result in the shot-pellets not spreading, but entering the body as a single mass. From greater distances, the pellets will spread and cover a greater area of the body surface; so, generally speaking, the greater the area of spread, the farther away the shotgun was fired.

Primer residues will also be left on the clothes of both victim and murderer, as well as on the murderer's hands, in the latter case especially from handguns. Chemical tests can be used to determine whether gunpowder residues are present on skin or clothes. These tests are now highly specific, but for a long time the dermal nitrate test was widely used. Unfortunately, this test was eventually shown to be very undiscriminating and positive results could be obtained from skin or clothes that had been contaminated with tobacco, fertilizers, cosmetics or urine. The test is no longer used, but it is salutary to reflect that forensic science is constantly developing and any current tests must be looked upon as potentially fallible, a fact many forensic scientists, lawyers and police officers have difficulty in accepting.

It is often the case that doubt exists as to whether a fatal shooting is murder or suicide. Clearly, if the distance from which a bullet was shot was greater than arms length, as can be shown from the presence or absence of skin discoloration, then foul play must be suspected. Other cases are more complicated and require a more searching investigation. The following case makes the point.

A young woman was found dead indoors, with a wound in her neck. Beside her was a .22 semi-automatic rifle. Her two children, as well as her mother and father, were also found dead. It seemed that the woman had murdered her family, then shot herself. However, a search of the premises resulted in the discovery of a silencer in the gun cupboard; it was wiped clean on the outside, but still had blood on the inside. If the silencer had been used in the murder, then it is obvious that the woman could not have shot herself, then gone to the cupboard, placed the silencer in it and, finally, returned to die beside the rifle. The woman's brother was the only surviving member of the family; he was arrested and convicted of murdering the five members of his family.

Another suicide-or-murder case took place in 1889 and has not, to this day, been finally resolved. Yet, like the case of King Ludwig, it is a case in which a small piece of evidence seems to have been overlooked. It is the case of Crown Prince Rudolf of Austria, who was found dead, together with his lover, Baroness Maria Vetsera, in the hunting lodge at Mayerling.

Prince Rudolf, son and heir of Emperor Franz Joseph, was depressed for personal and political reasons that need not concern us here. On January 28, 1889, the prince went to stay at Mayerling and was joined shortly afterwards by Maria. On the evening of January 29, the prince went to bed at about 9.00 pm, telling his friend, Count Hoyos, that he would see him at breakfast the next morning. Count Hoyos then retired to his apartments some short distance away

Shortly before 8.00 am the following morning, his valet announced that one of the guards wished to see him. The man said that the prince's servant, Loschek, was unable to wake the prince. Count Hoyos suggested that the prince must be sleeping very soundly, but was informed that His Imperial Highness had risen at 6.30 am and told Loschek to wake him up again at 7.30 am. Loschek had been knocking on the prince's door, first with his knuckles, then with a piece of wood, since 7.30 am, but there was no response. Count Hoyos, together with Prince Coburg, who had just arrived, rushed to the lodge and ordered that Prince Rudolf's bedroom door be burst open. Loschek went in, returning almost immediately to say that both Rudolf and Maria were lying dead in their beds. Rudolf was lying bent over on his bed, with a great pool of blood on the floor beneath him. The extraordinary conclusion was that he had suffered a haemorrhage as a result of taking cyanide administered to him by Maria – extraordinary, since the prince's forehead was smashed.

Eventually the news was broadcast that Prince Rudolf had died of heart failure. Even after it became obvious that both he and Maria had died of gunshot wounds, Emperor Franz Joseph flew into a rage when the matter was mentioned, insisting that "the woman" had poisoned him. Rudolf and Maria had written several letters of farewell; and Rudolf put his affairs in order, all of which pointed to suicide.

Franz Joseph's insistence that Maria poisoned Rudolf then killed herself was further undermined by the discovery that she died several hours before he did. The condition of the two bodies made that abundantly clear. The story that has survived to this day is that Rudolf killed Maria in a death-pact, then killed himself some time between 6.30 am and 7.30 am on January 30. When Rudolf gave his last orders to Loschek, Maria was already dead.

Later, rumours began to circulate that the prince had not committed suicide, but had been murdered. Four lines of evidence seem to support this idea. The first is a report by Prince Reuss concerning a conversation with Cardinal Galimberti, in which the latter expressed his firm belief that Rudolf had been murdered. The second is the fact that Rudolf was alive for several hours after the death of Maria, a scenario that is not consistent with a death pact. The third line of evidence is the fact that the grandson of the then Austrian Prime Minister, Count Taaffe, had seen documentary evidence that convinced him that Rudolf was murdered. The fourth piece of evidence is the published account of Countess Listowel, who wrote that an unnamed grandson of Rudolf had told her that his grandfather had been murdered.

The most interesting piece of direct evidence, which was alluded to by Cardinal Galimberti, is the fact that Rudolf's right temple and the right side of his forehead had been smashed and that the bullet hole at the lower left side of his head was very small. Galimberti knew enough about firearms to question whether this was possible, and suggested that the alleged exit hole was, in fact, the entry hole, indicating that the prince had been shot from behind, which in turn indicated that his death could not have been suicide.

This possibility was dismissed by Count Lonyay, who pointed out that the Austro-Hungarian service revolver, like the one that allegedly killed Rudolf, used 12.5 mm lead bullets without a steel casing. Therefore, he argued, it would spread on impact with the bone, causing a large wound and extensive damage.

All this is perfectly reasonable, but one point has been overlooked. If it is, indeed, true that the bullets used in the revolver were soft-nosed ones that would spread on impact, causing such frightful damage, *then the exit wound, too, would have been very large.* But it was not. This suggests that the bullet, in all likelihood, entered the head from the back lower left side and emerged through the forehead. The idea that Prince Rudolf was murdered is very probably true.

CHAPTER SIX

POISON

The leperous distilment; whose effect
Holds such an enmity with blood of man
William Shakespeare
Hamlet *Act I, Scene V*

I was standing in a corner of a small room in which a murder had been committed; at least, there was the dead body of a woman lying there and the police were treating it as a case of murder. But, as it turned out, it was not murder. As we – forensic scientists and police officers – were working, I saw one of the policemen bend over and pick something up from the floor. He then bent over again and picked something else up. His face was serious and concerned. The pathologist asked him what he had found and the officer held up a cup and a saucer, which had been lying beside the body. Instantly, the suspicion arose in the minds of all who were gathered there that this might be a case of suicide by poison, a suspicion confirmed by later investigations.

For some reason, poison has an enduring fascination. Most people seem to be both repelled and attracted by it, possibly because it is deadly and, at the same time, discreet. Its effects, unlike those of more obvious weapons, are not messy and unpleasant. And yet, for all its fascination and despite all we know about poison, we have no adequate definition for the word.

I remember reading a newspaper article about a man who drank six pints of water in twenty minutes, then fell down in fits and died. Imagine what one would have thought, had the substance been anything other than water. Most people, including most scientists, would probably have wondered whether a substance that could kill at such a high dose would be safe to take at all. "If six pints of that stuff can kill instantaneously, just think what the cumulative effect of swallowing a few drops every day would be," people might have said.

The point of this little anecdote is that almost any substance could be a poison – almost any substance can harm or kill, depending on the exact circumstances. Essentially, the toxicity of a substance depends on how much of it is ingested and over what period of time. Also, the age and weight of a person, as well as their state of health, would determine whether or not a swallowed substance would cause death or injury. Of course, what we usually mean when we say that something is a poison is that a very small quantity of it can kill. But what is a small quantity? Cyanide, normally considered a deadly poison, can be taken in small doses – less than 300 milligrams – with impunity. Indeed, small amounts of cyanide are essential for good health, since without it the body cannot synthesize vitamin B12. The "poison" is present in the seeds of many fruits, notably apples, peaches, cherries, plums and, famously, in almonds.

None of this means that cyanide is not a deadly poison; my point here is to illustrate the fact that even the deadliest poisons are deadly only above a certain dose. A man obsessed with his health lived for three years on nothing but carrot juice, which he was told was healthy. One day he was found dead, his skin a bright orange colour. Carrots are certainly good for you, but not in superabundance and to the exclusion of all else. The precise circumstances of this case made carrot juice a poison.

The point that must be remembered is that *anything* can be a poison. In a British trial in 2000, a nurse was accused of murdering her non-diabetic husband by injecting a huge amount of insulin into his leg. He was paraplegic and could not feel the prick of the needle. In 2008, another nurse was convicted of murdering four non-diabetic elderly patients who were "a burden on nursing staff" by administering large doses of insulin. In 2011, one of the worst cases of insulin poisoning came to light in a UK hospital, with the deaths of at least three patients with "unexplained low blood sugar levels". Investigations uncovered a batch of around 36 saline ampoules that had been contaminated with insulin and it was believed that at least eleven other patients had been affected but not seriously harmed. So far, the culprit has not been caught.

The use of poisons as a means of murder has a long history. Cyanide, present in laurel leaves, has been used to dispose of people since ancient times. The Empress Livia murdered her husband, Augustus Caesar, by soaking some figs in the poison before serving them. Yet it is interesting to relate that the Romans did not regard poisons as being wholly evil, for Pliny the Elder wrote that poisons, correctly used, were a good thing, because they could relieve men of the burden of living, when life became unbearable.

In forensic work, the questions that usually arise are: "Was the victim poisoned?"; "What was the poison used?"; and "Was it accident, suicide or murder?".

The first and second questions are, of course, linked. The suspicion that someone might have been poisoned arises from the general circumstances of the case, as we saw in the example at the beginning of this chapter. Once poisoning is suspected, scientific techniques can be brought to bear, but the precise technique will again depend on the circumstances. Sometimes suspicion of poisoning can arise without one having any idea what the poison may have been; no-one knew, initially, what poison the woman in the above case took. In such cases, screening for a wide range of possible poisons must be done. This is a laborious and time-consuming process.

TOXICOLOGICAL INVESTIGATION

More often, a particular poison is suspected. The first step in a toxicological investigation is, of course, the taking of a sample from the deceased, or, in a case of non-fatal poisoning, from the living victim. Samples from the latter are usually much easier to analyse, since the complicating factor of new decomposition products is absent. Before any actual analysis can be made, the sample must be treated in various ways in order to extract and purify the poison in it. The amounts of poison present in a sample are usually very small, since what we call a poison is something that kills at very low doses. Depending on their nature, poisons can be extracted using organic solvents, such as ether or chloroform, or by the use of solid silica absorbents.

Chromatography, a technique that separates the various compounds in a sample, is used as a means of identification. There are two main kinds of chromatography that are commonly used for toxicological screening, thin-layer chromatography and gas chromatography, each of which is used for a special purpose. However, each is based on the fact that different compounds will pass through an absorbent medium at different rates, which depend on their different chemical compositions. In gas chromatography, in which the test substance is made volatile by heating, the molecules from the various compounds will move through a long narrow tube of glass or steel, toward a detection device that "recognizes" the specific electrical impulses emanating from each compound. A recorder will then produce a chart plotting the concentration of each compound as a peak, against time. The time at which a compound appears – the so-called retention time – is an attribute of that compound; the height of the peak will reveal how much of it is present.

Nowadays, chromatography is coupled with a technique known as mass spectrometry. The mass spectrometer is a machine that bombards the molecules emerging from the chromatograph with electrons. This treatment ionizes the compound (i.e. breaks them up into electrically charged ions); and produces a spectrum, which is essentially an identification of each ion according to its atomic mass. The reason chromatography and mass spectrometry are used together in this way is that chromatography is not a certain method of identifying compounds. This is because the retention time, while an attribute of that compound, may be shared by other compounds as well. Essentially, it is a crude sorting process, after which the fine-tuning of the mass spectrometer can be brought into play. Actual identification is made by comparing the results obtained with control results from samples of known chemical composition.

Another identification technique is immunoassay, in which the fact that a foreign compound (an antigen) entering the blood will cause the body to produce a defensive antibody, can be exploited to good effect. Antibodies will bind on to antigens; the resultant "clumps" are then consumed by the white blood cells. In many cases, the antibody-antigen complex (the antigen here being a particular poison) can be identified. This technique was initially developed as an aid to blood typing, before being extended to drug detection particularly in blood and urine samples. Several assay kits for specific drugs are now commercially available, and new techniques, such as the enzyme-multiplied immunoassay technique or EMIT, are becoming very popular because they are usually highly sensitive and give a very fast result.

At this point it has to be said that, while the sophisticated techniques used in forensic science are very effective in identifying the many known poisons, the chemical make-up of most poisons is unknown. So many poisons occur naturally in plants and fungi, and even in animals, that cases of poisoning can occur, but cannot be identified as such. In Britain, a potential new source of poisons lies in the seeds that are often used as necklace beads among the Asian population. There are cases in which poisoning, either deliberate or accidental, by such seeds has been suspected, but about which nothing could be done.

ADMINISTERING THE POISON

The problem of how the poison was administered is often of some importance, since it may be relevant to the question of whether one is dealing with a case of accident, suicide or murder. Particular poisons are associated with particular forms of administration. For example, cyanide and arsenic are usually ingested through the mouth, while substances like heroin are often injected into the blood. Some poisons, such as snake venoms, are not poisonous unless they enter the body through the bloodstream. These venoms, being proteins, are simply digested in the stomach, if taken by mouth.

However, things may not be as simple as they seem. Josephus, the Jewish historian, records that "a noxious compound of the poison of asps and the secretions of other reptiles" was held in readiness to poison King Herod the Great in case the main poison did not work. (The plot failed.) This story, like various others from antiquity, suggests that it was believed that the venom of a snake could kill if swallowed – a belief that may have been well-founded, since any cut or graze inside the mouth would have allowed the poison to work. In any case, we know that the spitting cobra's venom enters the body through the mucous membranes of the eyes; and there is no reason to suppose that the venom of the asp (the Egyptian cobra) cannot enter through the mucous membranes of the mouth. In my view, it is very likely that these ancients knew what they were doing, since the art of the poisoner was highly developed and prized in those far-off days.

Drugs such as cocaine and nicotine – the latter as snuff – are often "snorted" into the nose, where they enter the body through the nasal membranes. Amphetamines are sometimes taken in the same way. More familiarly, eye and nose drops function in the same manner. Related to this route of entry into the body is the inhalation of substances, which reach the lungs and pass from there into the blood vessels.

Injecting drugs, such as heroin, directly into the bloodstream is a very effective means of introducing the poison into the system. Toxicologists use the term "bioavailability" to describe the percentage of the drug taken that will cause a physiological effect. Thus, the bioavailability of cannabis resin is very low if swallowed, since it is not easily absorbed at any point of the digestive tract. On the other hand, the bioavailability of injected heroin is effectively one hundred per cent, as are effectively all substances taken intravenously. Heroin taken orally, however, has a bioavailability of zero, to all intents and purposes.

Most poisons are, of course, taken by mouth. A major factor controlling the bioavailability of poisons taken orally is the metabolic action of the liver. When drugs are absorbed in the intestine, they pass into the blood vessels that carry blood to the liver, where they, like all other food, are metabolized for energy. It is in the liver that many potentially poisonous compounds are deactivated. So, although drugs may be absorbed in the digestive tract, their bioavailability is often drastically reduced by the liver.

Before the formidable barrier of the liver is reached, however, the first barrier – the poison's capability of being absorbed into the bloodstream in the first place – must be overcome. This readiness to be absorbed may be high or low. Much depends on the chemical composition and structure of the substance; we have already seen that an oily substance like cannabis resin is but poorly absorbed. It is for reasons such as these that different poisons are introduced into the body in different ways – the way chosen being the one that gives the substance its greatest bioavailability.

It is clear that the bioavailability of a drug or poison can be an important legal matter. If the substance ingested was of such a dose and taken in such a manner that it caused hardly any harm to the person to whom it was administered, then these points would be significant in a court of law. As we have seen in our discussion of techniques above, the forensic toxicologist can determine not only the type of poison used, but its dosage as well.

Metabolic activity usually results in the deactivation of poisons and their excretion. The excreted products are metabolites – substances produced by the breakdown or transformation of the ingested compounds – and are passed out of the body, usually in the urine. Some substances – heroin is a good example – are metabolized so quickly that the toxicologist in most cases has to detect the metabolites, rather than the drug itself. Therefore, a knowledge of the biotransformation (as it is called) of drugs is essential when attempting to determine the type and dosage of the poison. Sometimes, a drug is metabolized and its metabolites excreted so quickly that toxicological investigation may not succeed.

However, some compounds and their metabolites are retained in the body for an unusually long time, as compared with most drugs. Cannabis metabolites are an example of compounds that can be retained within the body for a long time – sometimes up to several weeks – facilitating the work of the toxicologist, or at least prolonging the time after ingestion and during which the search for signs of the drug is still likely to be fruitful. Some poisons are "excreted" simply by removing them into the hair or fingernails, where they can do little damage.

The toxicological facts and methods discussed above allow toxicologists to present much useful information on the course of events in a case of suspected murder or suicide. As we saw in the last chapter, people committing suicide behave in a fairly predictable way. In the case of suicide by poison, the "victim" usually removes his spectacles before swallowing the fatal draught. Murderers may not be aware of this and will fail to remove their victims' spectacles. Although this matter of behaviour may not be a scientific fact of toxicology, it suggests that a case may be one of murder, not suicide, and so should be followed up with a toxicological investigation.

INTERPRETING THE EVIDENCE

The toxicological techniques at the disposal of the forensic scientist may be very refined, but, in the end, it is the interpretation that matters. Let us say that the results show that such-and-such a poison at such-and-such a dose was used. What does it all mean? Perhaps it can be concluded that the kind and dosage of drug were such as to cause death. But what is the dosage that causes death? In fact, there are no hard-and-fast facts or figures. Some people are able to tolerate much higher levels of certain poisons, depending on their general medical condition, their size and weight and even their mental state. A sick person of slight build will succumb to much lower doses of a drug that would have little or no effect on a larger and healthier person.

Nor are such things the only potentially misleading factors. The concentration of a poison or a metabolite will vary according to where in the circulatory system a blood sample was taken. More confusing still may be samples from the urine, which can vary widely from person to person and, in the same person, according to their state of hydration. The amount of urine excreted is affected by how much water a person may have drunk, affecting the concentration of the drug in it. Coffee and tea are diuretics, which will result in larger volumes of urine. Some metabolites that could have arisen from poisonous compounds may equally well have had another, more innocent, origin. Decomposition products of a dead body may include compounds that confuse the issue.

This point was made during the trial of Dr Harold Shipman, described as Britain's most prolific serial killer. Shipman was put on trial for the murder of fifteen elderly women, whom he injected with overdoses of morphine and diamorphine, the main active component of heroin. Traces of the drugs were found in the bodies of the women, but the defence lawyers claimed that the drug levels could not be accurately ascertained, since the bodies were decomposed and, therefore, the results could easily have been misleading. However, morphine is a stable compound, which does not decompose very quickly after the death of the victim, although it is true to say that it is not known for certain how soon such decomposition will occur and to what extent and under what conditions. Although Dr Shipman's guilt appears to have been established on the basis of other evidence, including the forgery of the will of one of his victims, the point raised by the defence was, toxicologically, a valid one.

In view of all these sources of error, the interpretation of toxicological results must be made with extreme care. The toxicologist may need to be supplied with information about a victim's state of health, or the nature of his last meal. The place from which a blood sample was taken must be chosen carefully and notice of it taken when interpreting the results. If a dead body is infested with maggots, it is safer to conduct tests on the tissues of the maggots, since such living tissues would yield more reliable results; however, it must be borne in mind that maggots will concentrate certain substances in their tissues at higher levels than those obtained in the human tissues and this relationship should be understood and taken into account.

Toxicity curves – graphs plotting the dosage against the effect on the person – have been drawn up and may be used as a general guide. In principle, such a curve will rise as a shallow incline, then become much steeper, then, finally, level off. At the lower dosages, where the incline is shallow, the poison may have little or no effect. At the lower parts of the steep incline symptoms of mild toxicity will appear; whereas at the higher parts, symptoms of severe poisoning will appear. Finally, death will take place at the levelling-off phase, where, naturally, no further symptoms of ill-health will appear!

Poisoning as a means of suicide appears to be much commoner than it is as a means of murder. At least, that is what the figures appear to suggest. However, one must not forget that those poison murders that have been discovered are, by definition, the only ones that have come to light, if that is not a tautology. The fact that some poison murders are discovered only by sheer luck or accident suggests that many more such murders are committed than is generally supposed.

Accidents at work and at home, or medical error or malpractice, are other causes, the latter being, thankfully, very rare.

We will now look at the various common poisons and their effects on people. Chemists classify poisons according to their chemical composition, but here we will group them together on a functional basis, i.e. on the basis of the way they act on the human body.

CYANIDE

Most poisons act systemically; in other words, they act by adversely affecting the biochemical processes of the cells. Other poisons act in other ways, and we will consider these at the end of the chapter.

Some of the commonest poisons are called gaseous, or volatile, poisons, although they are not always taken in that form. We have already met one of these – cyanide.

What is called cyanide in the vernacular can be one of two chemical compounds: hydrocyanic, or Prussic, acid (HCN); and one of its salts, usually potassium cyanide (KCN). The acid is by far the more fast-acting of the two. In one case a laboratory technician was seen by one of his colleagues to raise a bottle of twenty-five per cent Prussic acid and literally drop dead before he could place the bottle down. By contrast, potassium cyanide may take up to twenty minutes to kill and, judging by the accounts of those who have witnessed such deaths, it is a horribly painful way to die.

Both the acid and the salt kill by blocking the mechanism that enables the cells to take up oxygen from the blood. A consequence of this is the flushed appearance of the victims, since the haemoglobin in the red blood cells remains saturated with oxygen, forming the normally unstable oxyhaemoglobin. (Its normal instability is what allows it to release the oxygen into the cells.) Oxyhaemoglobin has a bright scarlet colour, hence the red blotches on the skin of those who die of cyanide poisoning.

It is generally believed that Rasputin, who exerted such a strong influence over the Russian Imperial Family through his alleged ability to cure the Czarevitch, was given large amounts of cyanide, but refused to die, until he was shot several times, brutally beaten and thrown into the River Neva to drown. The dosage of cyanide given to Rasputin was said to have been one ounce – ninety-five times the lethal dose. It is said that Rasputin may have suffered from gastritis – in fact, this is an unsubstantiated rumour – a condition that slows down absorption of the poison somewhat, but this cannot possibly have prevented his death from an ounce of cyanide.

As we have seen, cyanide poisoning causes a livid colour to develop in the skin, and a smell of bitter almonds emanates from the mouth and from the body when it is opened during the post-mortem. Yet when Rasputin's daughter, Maria, saw his body after it had been recovered from the river, she wrote a description of it, but made no comment on the lividity of her father's appearance or the tell-tale odour, neither of which she could have failed to perceive had they been present. Yet the belief persists that Rasputin could resist the toxicity of cyanide, unlike all other human beings and animals. It is simply not credible that Rasputin was given cyanide; he died of his bullet wounds.

Two of Rasputin's murderers, Prince Felix Yusupov and Vladimir Purishkevitch, wrote accounts of what happened. Both agreed that the drinks and cakes at the party in Felix's home, the Moika Palace, contained cyanide. In view of what we know about cyanide, they must have been either lying or mistaken. The latter is the more charitable view, and there is some evidence to support it. Dr Lazovert was the conspirator who was said to have brought the poison and laced the drinks and cakes, but the doctor may have bungled the job. Felix wrote that he gave Dr Lazovert "a box" containing the poison. This implies that the cyanide was not in an airtight container, which is how it should be kept, otherwise it begins to decompose into a carbonate. When partially decomposed cyanide is swallowed, the carbonate, which has a corrosive and irritating effect on the stomach, usually induces vomiting in the victim, yet Rasputin showed no such symptoms. However, gastritis might have prevented the appearance of such symptoms in the relatively short period of time before he was shot.

It is interesting that Maria Rasputina believed that Dr Lazovert did not administer cyanide, but an opiate. This would explain Rasputin's drowsiness, as reported by Felix. Even more interesting is the rumour that circulated after Dr Lazovert's death, that the doctor made a death-bed confession, saying that he lost his nerve at the last minute and substituted a harmless powder. No-one has been able to trace this statement to its source, although on one occasion I thought the truth was about to be revealed. A letter appeared in *The Daily Telegraph*, saying that Dr Lazovert had told the actor, Christopher Lee, who played Rasputin in a film, that he never put cyanide in the cakes, and that Mr Lee had told the author of the letter this story. Intrigued, I wrote to Mr Lee, asking him to confirm this, but he telephoned back, saying that he had never met Dr Lazovert and had not even heard of the writer of the letter. Documentary evidence can be very misleading, as we shall see in the chapter "Words and Images".

CARBON DIOXIDE
Carbon dioxide (CO_2) is a natural gas, present at a concentration of about

0.03 per cent in the atmosphere. It is often present in industrial sites and in places where explosions or fires have just occurred. Its concentration in the lungs prevents the expulsion of CO_2 from the body, so that respiration is hindered. Concentrations as low as three per cent in the atmosphere will cause symptoms to appear – dizziness headaches and general weakness. Concentrations of over twenty-five per cent are fatal.

A case from the Sudan demonstrates how easily lethal concentrations can build up. In 1954, in the town of Kosti, three hundred cotton growers were locked in a room at 9.30 pm after some riots. The room measured about sixty feet by twenty feet and there was no room for anyone to lie down. The windows and doors were closed to prevent the men from escaping. By the time the room was opened at 5.30 am the following morning, one hundred and eighty-nine men had died as a result of the accumulation of CO2, as well as from heat and exhaustion.

Carbon monoxide (CO) is even more deadly, since it will bind on to haemoglobin to form the very stable compound, carboxyhaemoglobin. This means that oxygen can no longer be carried by the blood, since the site on the haemoglobin molecule at which the oxygen is normally bound is now permanently occupied by the CO.

Death by CO poisoning is a common form of suicide. The victim usually places his head in an oven or connects a tube from the car exhaust to the inside of the car and lies back. Death will follow soon afterward. Interestingly, people who commit suicide in this way take some trouble to lie comfortably, often arranging cushions and blankets before lying upon them to die. These facts are often useful in cases in which a death looks like suicide, but which it is suspected may be murder. In one case, a woman was found dead with her head lying in a greasy tray inside an oven. This was a highly uncomfortable and atypical position for a suicide and so murder was suspected. Tests showed that no carboxyhaemoglobin had formed in the blood; also, a mark on the woman's neck suggested death by strangulation, a suspicion that was later confirmed by the husband's confession that he had killed her.

The symptoms of CO poisoning during life are surprisingly similar to alcoholic drunkenness and it is not unknown for drivers to be overcome by CO if the gas is leaking into the interior of the car because of a faulty car-floor. People accused of drunkenness will sometimes plead CO-poisoning as a cause of their condition, but testing for the presence of carboxyhaemoglobin in the blood soon reveals the truth. One of the post-mortem signs is the development of a bright red tinting, which is remarkably similar to the colour changes due to cyanide poisoning. The absence of the distinctive bitter almonds smell, however, easily distinguishes the two.

The gross changes or odours that suggest that a particular poison was used are not the only indications; specific chemical tests are conducted on the tissues to confirm the original identification. For example, suspicion of cyanide poisoning can be confirmed by treating the tissues with silver nitrate, which produces a white precipitate of silver cyanide, which is insoluble in nitric acid. Further treatment of the precipitate with ammonium sulphide and ferric chloride will result in the production of a scarlet solution. Similar simple chemical tests are available for other poisons and chromatography or mass spectrometry need not always be used.

CHLOROFORM

A very well known, but largely misunderstood, volatile substance is chloroform. Although it is not often used as a poison, it is a very dangerous substance indeed, especially in view of the way it seems to be treated in whodunit films. Chloroform, far from inducing a person to sleep as soon as a chloroform-soaked handkerchief is placed over the nose and mouth, will actually make the victim struggle and, if asleep to begin with, to wake up. The convulsions caused by chloroform inhalation seem never to be shown in such films; the victim usually goes limp almost as soon as the handkerchief is clamped over his mouth. Although the poison will take effect in due course, it takes many minutes for it to do so. The usual picture of the victim succumbing immediately is not only misleading, but dangerous, since the point at which the victim relaxes is not far from the point of death due to paralysis of the heart muscles.

ALCOHOL

Probably the most easily available of all poisons is alcohol. To call alcohol a "poison" is not an exaggeration, since high doses can cause death within a few hours, or less. It is humanity's familiarity with, and use of, alcohol for most of its recorded history that allows us to think of it as a much more innocuous substance that it really is. Alcohol blood levels of over 40 basis points, or 400mg% (400 milligrams per 100 millilitres of blood), are usually fatal, but much lower doses can also be fatal, especially in young people. There is an entire class of chemical compounds called alcohols, but in the vernacular sense, alcohol means ethyl alcohol, or ethanol (C_2H_5OH).

Alcohol is a depressant, adversely affecting the function of the central nervous system. It is a substance that is absorbed very quickly into the bloodstream (within minutes), but it has the unfortunate attribute of taking much longer to leave the system. In other words, soon after the drink is swallowed, the blood alcohol level rises sharply to a peak, but the level subsequently declines slowly. The following graph expresses these facts visually:

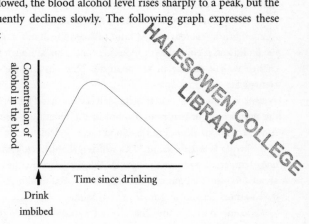

Alcohol consumed on an empty stomach will be absorbed much faster than alcohol taken with or after a meal. A high concentration of alcohol is absorbed more quickly than a similar dose at a lower concentration. Personal differences and medical conditions may also affect the rate of absorption.

Alcohol is removed in two ways: most of it – well over ninety per cent – is metabolized in the liver into acetic acid, which itself is then oxidized into carbon dioxide and water. About two per cent to five per cent is excreted in the urine. Some alcohol is released into the lungs when the blood carrying it reaches the alveoli, or small sacs within the lung. This is why alcohol can be detected in the breath and how it can be detected there by the various "breathalysers" used to establish whether a driver is in charge of a car after having drunk too much alcohol to make it safe for him to drive. Since it is impracticable, as well as being time-consuming and expensive, to take a blood sample from a driver who is suspected of being over the limit, the alcohol in the driver's breath is measured instead. Since there is a relationship between the amount of alcohol in the breath and the amount of alcohol in the blood, the measured breath alcohol makes it possible for the blood alcohol level to be calculated.

Now there is a fixed relationship between the amount of gas in a liquid and its amount in the air above it. Molecules of a substance will constantly move out of the liquid into the air and back again. Soon, a point will be reached when the number of molecules entering the liquid will equal the number of molecules leaving it. This is the point of equilibrium, which will differ in different temperatures, but which is fixed for a given temperature. At equilibrium, there is a fixed ratio between the number of molecules in the liquid and the number of molecules in the air, although that ratio will change with a change in temperature. This physical law is known as Henry's Law.

Applying this to the matter in hand, it is clear that if the temperature and the number of molecules of alcohol in the breath are known, then the concentration of alcohol in the blood can be calculated. The temperature of the breath is usually about 34°C, which is slightly below the normal core body temperature of most people. At this temperature, the equilibrium is about 2100 to 1; or, put another way, every 2100 millilitres of breath will have as much alcohol as 1 millilitre of blood.

We come now to how the alcohol concentration in the breath is measured. Most "breathalysers" contain potassium dichromate, with silver nitrate as a catalyst (a substance that speeds up a reaction, but which does not take part in the reaction and remains unchanged after it.) The alcohol in the breath reacts with the potassium dichromate. During this reaction, two molecules of potassium dichromate are removed for every three molecules of alcohol.

Potassium dichromate absorbs light in a certain part of the spectrum. The amount of light absorbed is proportional to the concentration of the absorbing substance (in this case, potassium dichromate), as expressed in the following formula:

$$A = kc$$

Where, A is the quantity of light absorbed, k is a constant and c is the concentration of the absorbing material. Of course, the formula can be expressed slightly differently, as follows:

$$c = \frac{A}{k}$$

From which the concentration can be easily calculated. Desktop breath analysers are generally chemical-free and instead use infrared spectro-photometer and fuel cell detector technology to achieve the same result, a measure of the concentration of alcohol in the breath sample.

The legal upper limit of blood alcohol for drivers in Britain is 80mg%, although it is very likely that it will be lowered. In the United States, the legal limit varies from state to state, some having very high, others very low, limits. Some states do not have a legal limit at all and deal with drink/driving offences on a case-by-case system.

The forensic interest in alcohol is not limited to drink/driving offences and outright fatalities of the kind described above. Drunkenness and chronic alcohol poisoning are two other aspects of the problem of alcohol abuse. It is worth mentioning that drunkenness is not necessarily a reflection of the amount of alcohol consumed in a population, but it is related to the way it is consumed and at what times. Thus, more alcohol is consumed per individual per year in France than in Britain, but there is far more drunkenness in Britain than there is in France.

Drunkenness, in itself, is not an offence, but it becomes so when the person in question is in charge of a vehicle – car, or ship or aeroplane – or in charge of children. Causing damage or injury while drunk is no defence under the law, although killing when in such an inebriated condition that the perpetrator did not know what he was doing will usually result in a conviction for manslaughter, rather than murder.

It is not very easy to define the term "drunk". Although establishing the blood alcohol level of a person is quite straightforward, having a certain high concentration in the blood does not necessarily mean that that person is drunk, since some people have a greater tolerance for alcohol than do others. Behavioural criteria, such as the ability to stand on one leg without falling over, the presence or absence of jerky movements of the eye and so forth, are good general guidelines, but they should be interpreted with care, since the person in question, due to ill-health for example, may have the "wrong" behavioural responses, even though he is not drunk.

Chronic alcoholism is a much more common condition than is generally supposed. It is expressed as a general degeneration, both physical and mental. Loss of weight and appetite, together with "fatty degeneration" of the muscles, including the heart muscles, is a common feature. The liver, too, degenerates. In some cases, inflammation and degeneration of the peripheral nerves may take place.

Alcohols other than ethyl alcohol can be abused, and people are known to drink methylated spirits, based on methyl alcohol, or methanol (CH_3OH). Even the added purple colouring does not seem to put off some hardened drinkers of this extremely noxious preparation, which includes acetone, bone oil, paraffin and various aldehydes, as well as methanol. It is sometimes filtered through charcoal, or even a very thick slice of bread, to remove the colouring, the resulting clear liquid being added to a cheap red wine and consumed. This concoction is known as "Flaming Lizzie" and is sold, quite illegally, in some pubs. Habitual abuse is often fatal.

STIMULANTS, EXCITANTS AND HALLUCINOGENS

Another dangerous group of systemic poisons are the stimulants, excitants and hallucinogens. Of these, the amphetamines are among the best known. A typical example is "speed", which produces feelings of alertness and confidence and makes the user feel more energetic. Unfortunately, abstinence from the drug results in the opposite effects: fatigue and depression. In some cases, it can cause heart attacks and strokes, leading to death. Continued use of the drug causes dependence and general ill-health, which leaves the sufferer prone to all kinds of illnesses.

"Ecstasy" is another amphetamine, revelling in the name of 3,4-methylene-dioxy-methamphetamine. As its common name implies, it produces euphoria – which may turn into hallucinations. Users of the drug experience sharper sensations – music sounds better and sympathy with others is increased. Unlike speed, ecstasy is not physically addictive, but its pleasant sensations fade after continued use. However, its use is often accompanied by feelings of nausea and the substance is known to cause damage to nerves in laboratory animal experiments and can cause depression in some people. Another problem is that ecstasy is often sold mixed with other drugs, the identity of which may not be known to the user. Youngsters taking the drug often drink alcohol as well, a habit that increases the probability of adverse reactions.

The infamous LSD (lysergic acid diethylamide) of the sixties is a very powerful hallucinogen. It is extremely dangerous, often causing mental delusions, such as paranoia and schizophrenia, as well as other psychological disorders like panic attacks and the impairment of judgement. It is not quite as common now as it used to be, but it is still in circulation. It is a synthetic drug, although the lysergic acid used in its manufacture is obtained from ergot, the poisonous fungus.

Many poisons are, of course, derived from plants, sometimes quite common ones. An example is atropine or belladona, which is derived from the Deadly Nightshade, *Atropa belladonna*. The dull purple flowers and shining black berries are well known. Children sometimes eat them, with serious, although usually non-fatal, results. Nevertheless, fatalities are known. The tell-tale berry skins are found in the stomach during the post-mortem. Child deaths due to accidental ingestion of this sort are, of course, extremely distressing, and they have a long history. In 1597, John Gerard wrote in his book *The Herball or Generall Historie of Plantes* that "...*three boies of Wisbich in the Ile of Ely, did eate of the pleasant and beautifull fruite hereof, two whereof died in lesse then 8 howers after they had eaten of them*". The third child survived after being induced to vomit. All children are apt to look upon bright, shiny berries as edible things; children with a special fondness for fruits are particularly likely to pick and eat wild berries, and it is wise for parents to be aware of this. Gerard in fact recommended that the "sleeping nightshade" be banished from gardens and places near the house because children and women alike to "*do oftentimes long and lust after things most vile and filthie; & much more after a berrie of a bright shining black colour, and of such great beautie, as it were able to allure any such to eate thereof*".

The symptoms of atropine poisoning include hoarseness and drying of the skin, as well as a condition akin to mania and rapid breathing, which is followed by coma and respiratory failure in severe cases. In mild cases, such as that reported in 2009 of a woman who ate six Deadly Nightshade berries thinking they were blueberries, symptoms may include blurry vision, lightning flashes, disorientation, loss of balance and feelings of anxiety and agitation that last for 24 hours.

Thorn-apple (*Datura stramonium*) and Henbane (*Hyoscyamus niger*) are the sources of hyoscine, a drug that is also found in Deadly Nightshade. It causes delirium and renders the user incapable of making reasoned judgements, often blurting out whatever is on their mind. It is this symptom that has led to it being used as a "truth drug", under the more familiar name of scopolamine. The most famous, and possibly one of the first, cases in which hyoscine was used to commit murder, was when the infamous Dr Crippen used it to dispose of his wife.

Cocaine is another plant drug, being derived from the dried leaves of Erythroxylum coca, a tropical plant grown in Africa, South America and

Southeast Asia. It is a powerful nerve stimulant and overdoses can kill by overstimulation of the heart. It is a most dangerous drug, which, when it does not kill, degrades the addict to such an extent that a description of the symptoms would be indelicate. In this, cocaine is similar to heroin and other narcotics. It is usually taken as a snuff.

More commonly these days, "crack" cocaine is used. This is produced by heating cocaine with baking soda in water. The resulting mixture is dried and is used for smoking. It is much more dangerous than cocaine. It is worth making the point here that one of the difficulties that sometimes arises in toxicological testing is that someone may pick up traces of cocaine and other drugs as a result of passive smoking, having been sitting in a room where someone else had been smoking the drug. This is common enough with tobacco smoking, but it applies to other drugs as well.

CONVULSANT POISONS

Allied to the stimulant drugs are the convulsant poisons, of which strychnine is one. It has an exceptionally bitter taste and is extracted from a number of Indian trees, notably *Strychnos nux-vomica*. It causes powerful convulsions and death follows from the paralysis of the medulla oblongata in the brain. The symptoms are not unlike those of tetanus, with a stiffening and rigidity of the muscles, although the two conditions can be separated by the fact that, in strychnine poisoning, the chest muscles are particularly prone to fixation, whereas in tetanus it is the jaw muscles that stiffen most noticeably. Also, in strychnine poisoning, almost any touch or sound will trigger a convulsive bout.

Although strychnine is not commonly used in murder cases, it is used as a rat and mole poison and is available to people whose profession involves the killing of vermin. This fact helped to solve an extraordinary murder case – extraordinary, because of the way it came to light and by the way the culprit gave the game away at almost every step.

On May 26, 1934, Inspector Dodson of the Lincolnshire police received the following letter:

Sir, have you ever heard of a wife poisoning her husband? Look further into the death (by heart failure) of Mr Major of Kirkby-on-Bain. Why did he complain of his food tasting nasty and throw it to a neighbour's dog, which has since died? Why did it stiffen so quickly? Why was he so jerky when dying? I myself have heard her threaten to poison him years ago. In the name of the law I beg you to analyse the contents of his stomach.

        signed,
                Fairplay

The identity of the writer has never been discovered. Arthur Major had, indeed, died two days earlier. The first thing Inspector Dodson did was to ask the doctor who issued the death certificate what was the cause of death. He was told that Major had been taken ill with violent convulsions and spasms and that the cause of death was epilepsy. Dodson then went to see Mrs Major, and he saw Major's body. As he was leaving, Mrs Major asked: "I am not suspicioned? I haven't done anything wrong", which was an odd comment from one who had not been accused of anything.

Major had died on May 24, and on the previous day, May 23, the wire-haired fox terrier of his neighbour, Mr Maltby, died after suffering severe muscular contractions. Maltby buried the dog in the garden. In order to look further into this strange coincidence, the dog's body was exhumed. At the same time the coroner ordered that the funeral of Mr Major be stopped. Post-mortems of both Major and the dog were carried out and organs from both were sent to Dr Roche Lynch, the Home Office toxicologist, for analysis.

Dr Lynch found fatal doses of strychnine in both bodies. Moreover, he concluded that Mr Major received two separate doses – one on May 22 and another on May 24, the day he died. Death could not have been suicide, since once the agony of the first dose had been experienced and passed, only a lunatic would take a second dose.

Police investigations revealed that the Majors' married life was far from happy. They quarrelled frequently and Mrs Major frequently accused him of drunkenness, idleness and of having affairs with other women. As proof of this last accusation, she produced letters from alleged lovers and from anonymous accusers, saying that Major was a lecher. She handed over all these letters to the police. One letter, sent directly to the Chief Constable of Lincoln, alleged that Mr Major was always drunk and was not safe on the road. Mrs Major also told the police that her husband had been trying to kill her by "putting something" in her tea.

Most significantly, Mrs Major had run up a number of debts and her husband had been getting concerned about this. She spent inordinate amounts of money on clothes for herself, although they were not an affluent couple. In exasperation, Mr Major decided to place a notice in the local newspaper, dissociating himself from her debts. On that day he died; and Mrs Major cancelled the notice submitted to the paper.

The police interviewed the couple's 15-year-old son, Lawrence. He told them that relations between his parents had deteriorated to such an extent that he and his mother slept at her father's house, although she still cooked her husband's meals. When asked about the events of May 24, the boy said that his father was sitting trembling in his chair and that his mother said that the corned beef in his meal had upset him. Major then began to foam at the mouth, saying that he was going to die, which he did at 10.40 pm.

Interestingly, Mrs Major continued to volunteer information to the police. She said that she had nothing to do with the preparation of his meals and insisted that her husband had bought the "tinned beef" himself. She was also at pains to impress upon the police that she had been staying with her father during the two weeks prior to her husband's death. This interested the police, but in a way that Mrs Major had, perhaps, not intended, for her father was a gamekeeper, who kept poison to kill vermin. When the police asked her whether she was aware of this fact her reply, astonishingly, was: "I did not know where he kept the poison. I never had any poison in my house and I did not know my husband died of strychnine poisoning." When the interviewing officer, Hugh Young, asked her why she mentioned strychnine, since he had never mentioned it, she merely said that she had made a mistake and that she was certain that her husband had died of the corned beef.

Time and again, she returned to the corned beef. She said that she had seen the tin on the pantry shelf, then she made another astonishing statement: "It was quite black. I thought at the time it was bad, but I did not tell my husband." She seemed utterly unaware that she was condemning herself. Further damning evidence followed: Mrs Major told the police officer that she did not send for the doctor when her husband was ill the second time, but gave no further explanation.

The police then interviewed young Lawrence again, this time with his mother present. He told the police that his mother had sent him to buy the corned beef, but his mother denied this, saying that it was his father who had sent him. Interestingly, the boy insisted that it was she who sent him.

There was one last interview for Hugh Young to carry out. He went to see Mr Brown, Mrs Major's father, and asked him whether his daughter knew that he kept poison. "Yes, mister," was his answer. He kept it in a locked box and kept the key with him at all times. Would his daughter have been able to open the box? "No, mister."

There was enough evidence against Mrs Major, who was arrested. But the question of where she obtained the poison was still unanswered. After the arrest, Young searched her bags and discovered a small key, wrapped in a piece of paper. He went to see Mr Brown again and asked him whether he had another key to his poison box. He said he had had one, but he had lost it some time ago. Young had put the key on his own key ring, which he held up to Mr Brown, asking him to select the one that fitted the box. He chose the one taken from his daughter's bag. It fitted the lock and opened the box.

Ethel Major was put on trial for murder, was convicted and hanged. For some reason, the jury had found her guilty but recommended mercy, a recommendation that was ignored by the judge.

Aconite is another convulsant of plant origin, being extracted from Monk's Hood (*Aconitum napellus*), also known as Wolf's Bane, because the poison was added to dead carcasses in order to kill wolves during Mediaeval times. It is one of the fastest acting poisons known and there is no known antidote; the Ancient Greeks called it the Queen of Poisons. Its symptoms are at first a tingling in the throat, followed by a severe burning sensation. The limbs grow weak, speech becomes confused and respiratory failure takes place. Monk's Hood should not be grown in gardens that small children are likely to frequent.

Other plants containing convulsant poisons include the yew tree (*Taxus baccata*), on which both the leaves and berries are deadly. Laburnum leaves and fruits are also poisonous. Foxgloves (*Digitalis purpurea*) are poisonous, although not as deadly as is sometimes stated. Other common poisonous plants worth mentioning are Marsh Marigolds, Bryony, Holly (berries), Juniper, Daphne, Snowberry, Privet, Honeysuckle, Wood Anemones and Greater Celandine. The bulbs of Daffodils and Hyacinths, as well as the roots of Bluebells, are also toxic.

A large number of species of mushrooms produce poisons that range from the stomach-upsetting kind to the instantly fatal. This is not the place for a full discussion of this topic, but it is worth mentioning a few of the most dangerous species, since they are the cause of some quite avoidable deaths. Fly Agaric (*Amanita muscaria*), the well-known red-capped toadstool with white "warts", is very poisonous; worse still is the Death Cap (*Amanita phalloides*), which is the classic poisonous mushroom, causing most mushroom-poisoning deaths. The closely-related *Amanita pantherina*, or Panther Cap, is also extremely poisonous.

Interestingly, there is a number of false beliefs relating to the toxicity of mushrooms. One is that poisonous mushrooms, if boiled, will be rendered harmless. This is true only of some species, but by no means all. Another is that mushrooms that exude "milk" when broken are poisonous; again this applies only to some species. Yet another dangerous belief is that all spring mushrooms are safe to eat; this is not true. One of the most delightful (were it not so dangerous) myths is the notion that mushrooms that do not cause a silver spoon to change colour while they are being cooked are safe to eat. Once again this is not true. Many of the Boletus mushrooms are good to eat, yet at least one species, Satan's Boletus (*Boletus satanas*), is very poisonous.

One particular mushroom, *Psilocybe semilanceata* (the Liberty Cap), is of some importance in forensic science, since it contains a hallucino-genic poison. In fact, something like eighty species of this genus produce delirium and hallucinations, but the Liberty Cap is by far the most common, and the most potent, of these so-called "magic" mushrooms.

A widespread and superficially logical belief is that anything that is eaten by wild animals is also suitable for human consumption. Unfortunately, it is not true. Snails and slugs will eat Death Caps quite happily; blackbirds will eat the poisonous (to us) berries of Barberry (*Berberis*) and stick insects will eat Privet leaves.

It may come as a surprise that ricin, a toxin found in the seed of the castor oil plant, is one of the deadliest. It is present in the hard shell of the seed but the oil itself does not contain any. One very daring murder was committed using ricin.

In September, 1978, Georgi Markov, an expatriate Bulgarian living in London and working for the BBC, was standing at a bus stop on Waterloo Bridge. He felt a sharp sting in his thigh and turned to find a man picking up his umbrella. Later that night, Markov's temperature rose to 104° F and he was having bouts of vomiting. He was admitted to hospital, where a small hole in his thigh was noticed by a doctor. Four days after he became ill, he died.

Georgi Markov was a political dissident and was considered dangerous by the then communist rulers of Bulgaria. He himself spoke of being poisoned by agents of the Bulgarian government, although only his wife, Annabel, took him seriously. After his death, however, a portion of skin around the hole in the thigh was sent to Porton Down, the government microbiological laboratory. A small pellet, capable of holding less than half a milligram of liquid, was found embedded in the skin. If the pellet had contained poison, it would have seeped out through the two small holes found in it. By a process of elimination, based partly on the poisons that would kill at as low a dosage as half a milligram, the scientists at Porton Down concluded that the poison was possibly ricin. A similar amount of ricin was injected experimentally into a pig, which later died, after showing much the same symptoms as did Markov. After the case was broadcast, reports about another Bulgarian dissident, this time in Paris, began to circulate. He, too, had died after feeling a sting in his leg while waiting for the Metro.

It seems very probable that the poison used to kill Markov was, indeed, ricin, but this has not been established beyond doubt. I have also heard that there was evidence that some other substances, such as extracts from Mistletoe (*Viscum album*), were used in the pellet.

TRANQUILLIZERS

The analgesic, hypnotic and narcotic drugs – all of which are "tranquillizers" of one sort or another – are among the commonest poisons used in suicides, simply because many of them are so easily available at home. Paracetamol is one of the commonest drugs at home and people have attempted to commit suicide by taking large overdoses of it. While very high doses will certainly kill – even five grams may be fatal in some cases – what is not generally known is that death is never very rapid and, if I may put it in a somewhat macabre fashion, it is not the method of choice for suicides. Two or three days of suffering usually elapse before death follows, due to liver failure.

Cases of fatal poisoning can be quite complex, death often being due to a number of combined causes. In one case a woman who had a history of suicide attempts was taken to hospital three and a half hours after she had taken an overdose of paracetamol, together with an antidepressant, prochlorperazine. At this point, the level of paracetamol in the blood was found to be 270 milligrams per litre, which is extremely high. In view of this, the doctor administered the antidote drug, acetylcysteine. Inexplicably, however, a ten-fold dose of the usual therapeutic amount was given, despite the fact that both the nurse and the pharmacist expressed reservations. Astonishingly, another ten-fold dose was administered three hours after the first. The woman died two hours later of cardiac arrest.

Unfortunately, the effect of high doses of acetylcysteine is unknown, although the results of experiments on dogs suggest that the drug is safe at high doses. The drug acts by protecting the liver from damage. In the above case, a paracetamol level of 270 milligrams per litre four hours after ingestion suggested that liver function was considerably impaired. At the post-mortem, the level of blood paracetamol was 102 milligrams per litre, indicating a drop of a little less than 170 milligrams per litre in eight hours. This result indicated that the deceased's liver had been severely damaged, since a normal liver would have reduced the blood level of paracetamol to about 70 milligrams per litre in that time.

The amount of prochlorperazine in the blood at post-mortem was 0.07 milligrams per litre. The normal amount to be expected in the blood when the drug is administered at the correct therapeutic doses is between 0.01-0.14 milligrams. The "half-life" of the drug, i.e. the time at which its original concentration will have been halved, is about six hours. This indicated that the highest level in the blood must have been in the region of 0.28 milligrams per litre – a dose well within the dangerous range.

So, what caused the woman's death? In fact, it is not possible to ascribe death to any one toxic cause. It would seem clear that death was caused by a drug overdose, which is what the court inquiry concluded, but it is impossible to be more specific.

Aspirin is another common analgesic; it can be fatal at doses of 25 grams. The Romanovs – a family with a rich supply of forensic conundrums – provide a possible example of how aspirin can be used to cause harm. As we saw in the chapter about identity, the Czarevitch suffered from some kind of bleeding disease. Aspirin can cause irritation of the stomach and, in some people, can cause bleeding, especially in those who are normally prone to such bleeding. Interestingly, there is evidence that Rasputin administered aspirin to the boy, whom he then "cured" by the simple means of withholding the drug, thus giving the impression that he could work miracles, which is what the Czarina and many others believed.

Among the hypnotic drugs, the barbiturates must rank as among the most dangerous. Often prescribed for tranquillizing purposes, they are frequently taken with alcoholic drinks, the latter enhancing the effect of the chosen drug while greatly reducing the amount needed for a fatal dose. A dose of one gram of some barbiturates can kill within minutes or even seconds, although most kinds usually take a few hours to take effect. Generally speaking, ten times the normal medicinal dose is fatal in most barbiturates – a dose that may not seem so high to the patient.

When an inquiry is treated as a suspicious death, the suspicion that poison may have been the cause of death sometime arises in the oddest of ways. In the case of Terence Armstrong, a six-month-old baby, poison was suspected only when what looked like the red skins of berries were found in the child's throat. Wondering how a baby of that age could have eaten poisonous berries by accident, the investigators decided to examine the "skins" more carefully. They proved to be the gelatine capsules used to contain various medicinal drugs. Toxicological analysis later confirmed this, demonstrating the presence of the barbiturate Seconal in the baby's body.

Both parents were eventually put on trial and the jury found the father guilty; he was condemned to death, but later reprieved. The mother was found not guilty, although she later admitted to having given the child Seconal to make him sleep.

The misuse of anti-depressants can sometimes be the cause of suspicion. In one case, a woman was found lying face down and half-suffocated in her house. The room she was in was in total disarray, with furniture thrown about and curtains drawn, although it was daytime. Her own clothes were torn off or dragged down and she had many bruises on her body.

Initially, it was thought that the woman had been assaulted by an intruder, but the truth emerged when it was discovered that her body contained an extremely high overdose of the anti-depressant Tofranil. Such high doses can cause great agitation and hallucinations.

NARCOTICS

We come now to the narcotic drugs, of which opium and its derivatives are the most familiar. A narcotic is a substance that dulls pain and reduces consciousness. Opium is the dried juice of the opium poppy (*Papaver somniferum*), which is the substance that goes into the pipe of the opium addict. Morphine is a constituent of it and heroin, or diamorphine (also known as diacetyl morphine), is derived from morphine. Other narcotic substances are found in the opium resin extracted from the plant's seeds – thebaine, narcotine, papaverine, and codeine, among others. The legitimate use of opium is the extraction of codeine and morphine, which are used medicinally. Laudanum, beloved of Victorian poisoners, is opium dissolved in alcohol.

The symptoms of opium poisoning are stupor – leading to coma – and heavy sweating. Sometimes the pulse beats very quickly and, as this is a condition caused specifically by thebaine, it indicates that the victim has been using (or has been given) a fairly crude extract, rather than refined heroin. Later, the pulse becomes slower, breathing also slows down and the body becomes cold. Some of these symptoms mimic those of certain kinds of haemorrhage, but the normal or lowered temperatures typical of opium poisoning distinguishes the latter from the former. The most distinctive symptom of opium poisoning is the contraction of the pupils to "pin-points". This symptom is so characteristic that it was used decisively and very dramatically in a trial in the United States during the late nineteenth century.

In 1892, Mrs Anna Buchanan, a former prostitute who was married to Dr Robert Buchanan, a New York doctor, fell seriously ill and died, according to the medical examiner, of a brain haemorrhage; but Mrs Buchanan's former lover (and pimp) did not believe that this was the true cause of death. After a long press campaign, the New York coroner ordered an exhumation and a toxicological investigation of the remains. The distinguished toxicologist, Professor Rudolf Witthaus, conducted the investigation. He found one-tenth of a grain of morphine, which, by back extrapolation, was equivalent to a fatal administered dose of two to three grains.

Dr Buchanan was put on trial for the murder of his wife. Evidence was presented that Buchanan ridiculed a man named Carlyle Harris, a medical student, as a bungler, because he had failed to put some drops of belladonna in his wife's eyes before poisoning her with an overdose of morphine. Belladonna, said Buchanan, would prevent the tell-tale "pin-pointing" of the eyes and, if Harris had thought of doing so, he would have got away with murder.

Unfortunately for Buchanan, a number of people remembered that he had put drops in his wife's eyes shortly before she died and it was decided to conduct a simple, though somewhat macabre, experiment in court. A cat was poisoned with morphine and, as it was dying, a few drops of belladonna were put in its eyes. Sure enough, the pupils did not contract, proving that Dr Buchanan did, in fact, know what he was doing. Both he and Harris were found guilty of murder and were sentenced to death, both dying in the electric chair at Sing Sing Prison.

Since no two preparations of heroin will have the same percentages of impurities, it is possible for the forensic toxicologist to determine whether two samples came from the same batch or not. Sometimes the difference between two samples may be obvious from the difference in colour alone, although chemical and chromatographic tests can be carried out to confirm that they are different. Also, heroin derived from opium poppies grown in different parts of the world may have different chemical characteristics. What are known as "cutting agents" can also help to distinguish between batches. A cutting agent is simply a substance, like sugar or a barbiturate, used to dilute the drug, or to camouflage it to obscure the fact that it is impure or lacking in a particular ingredient.

I am sometimes asked what is the fatal dose of morphine or heroin. Unfortunately, with these most dangerous of all drugs of abuse, it is not possible to give a simple answer. This is because habitual users may acquire a tolerance to the drug and can withstand much higher doses than can people who are new to it. In general, however, dosages of over two hundred milligrams are certainly lethal, although many people will succumb to much lower doses. Children are very susceptible to morphine; in one case a 6-year-old child fell into a coma and died five hours after being given 10.8 milligrams of morphine in error.

As with other drugs, the toxicologist must be aware that some source of morphine may be quite innocent. Small quantities of the drug may be found in poppy seeds used in baking and traces of it may appear in the urine. As we saw with cocaine, the detection of a drug on or in a person does not necessarily mean that foul play or abuse has taken place.

One of the commonest and oldest drugs is cannabis, the product of the plant *Cannabis sativa*. It can be met with as the dried leaves, stems or flowers, as a resin or as an oil. The fresh material smells like spearmint, albeit with a difference. It is often suggested these days that cannabis should be legalized, because it is less addictive than alcohol. This seems to say more about alcohol than about cannabis, although this is probably not a popular point of view!

Paralytic poisons include the toxin coniine found in the leaves of Hemlock (*Conium maculatum*), the poison with which Socrates was executed. Like other poisons in this group, it blocks neuromuscular action. Its effect is slow, which is what allowed Socrates to continue his famous discourse for some time after swallowing the fatal draught. Coniine first paralyses the legs, then the rest of the body. Death follows after respiratory failure.

Nicotine, most commonly met with in smoking, is not (smoking apart) a widely used poison, but fatalities occur with more frequency these days, since the purified toxin is used in various insecticides and pesticides. Three or four drops of pure nicotine will kill in a very short time. There are many other paralytic and related poisons, but they are rarely met with in forensic science.

OTHER POISONS

We have seen that most poisons are toxic as a result of their very precise biochemical effects. Other poisons, including some very well known ones, work by means of a general effect on the tissues as a whole. Nevertheless, some of these poisons have more specific actions, but, generally speaking, their ability to cause death is due to their more general effects.

Strong acids and alkalis, as well as the salts of heavy metals, have a corrosive or "burning" action on the tissues. Phenol (or carbolic acid) and related compounds are commonly used in disinfectant and antiseptic preparations for domestic use and, therefore, are among the most frequently used suicide poisons. Swallowing phenol results in a burning pain in the mouth and the digestive tract. Burn marks on the lips of suicides suggest that phenol or one of its close relatives has been used; the characteristic sweet smell of phenol is also a tell-tale sign. Death is due to the general destruction of the tissues, as well as the depressing activity the poison has on the central nervous system.

Lysol is the brand name for a range of disinfectant household cleaners, with a variety of formulations, but it can also be regarded as another poison in this group. Its effects are similar to those of phenol, although a recorded toxicological case reveals how dangerous these compounds can be. A man who kept a bottle of lysol in his hip pocket fell asleep and the bottle was accidentally broken. The liquid poured out, soaking his entire leg. The man died three-quarters of an hour later, although he had not swallowed any of the poison; the lysol was absorbed through his skin, especially through wounds or scratches, with the consequent effect on the nervous system that caused the man's death.

Oxalic acid is found in many preparations used for cleaning metal and leather and is used extensively in brass-polishing and in the restoration of old book bindings, among other things. It is said that many accidental deaths from oxalic acid poisoning are consequences of the fact that its crystals look very much like those of Epsom salts and are easily mistaken for them. Indeed, there are cases on record in which hospital patients were given oxalic acid in error, the nurse believing the contents of the package to be Epsom salts.

A secondary effect of oxalic acid poisoning is its depression of blood calcium levels. The resulting crystals of calcium oxalate are precipitated in the kidney tubules, a sure post-mortem indication of oxalic acid poisoning.

Oxalic acid is the toxin found in rhubarb leaves. It is not found in the stems, although, unaccountably, the belief has arisen that it is found in the stems, but that boiling neutralizes the poison. This has resulted in some people boiling the leaves and eating them as a vegetable, with fatal results.

The other main group of non-systemic poisons comprises those known as irritants. These substances cause great irritation to the stomach lining and, although they are not fast-acting, they include one of the most commonly used poisons in history – the semi-metal arsenic. Arsenic is obtained from its ore, arsenopyrite, which is a greyish-white mineral containing iron and sulphur, as well as arsenic. The white oxide is extracted by placing the ore in a container and heating it to a high temperature, when the vapour produced condenses on the walls of the vessel. In ancient times, extraction was a dangerous process, since the garlic-smelling vapour is extremely poisonous.

Arsenic produces much the same symptoms in whatever form – any of its salts or its vapour – it is administered. Vomiting, often blood-speckled, is one of the earliest symptoms, followed by intense stomach pains, diarrhoea, thirst, nausea, cramps, loss of weight, a feeling of burning and constriction in the throat, and melanosis (darkening) of the skin. These are symptoms of chronic poisoning, but in acute cases in which a large dose was ingested, the pulse grows weak and muscular convulsions may take place before death.

Arsenic was such a popular poison for so long partly because it is effectively tasteless, having only a very faint sweet taste, which can easily be masked by the flavours of the food in which it is administered. Nowadays, at least in developed countries, arsenic is less easily available, as it is no longer used in the manufacture of various dyes, although the twentieth century has seen some cases of murder and suicide using arsenic-containing weedkillers. In Africa and India, arsenic continues to be in use as a poison.

Arsenic has the distinction of being the first poison for which a diagnostic test was devised – the Marsh Test. This involves the addition of the suspected fluid (which may be an extract from tissues), mixed with a small quantity of sulphuric acid, to a piece of zinc. The reaction of the zinc and sulphuric acid will produce only hydrogen, but if arsenic is present in the solution, arsine, or arseniuretted hydrogen, will be given off. This gas can be detected by igniting it, then holding a piece of glass above the flame. If arsenic is present, it will be deposited on the glass.

Another useful attribute of arsenic is that it will remain detectable in the body long after death. This is because much of it is "excreted" by the simple expedient of stashing it away in parts of the body, such as the hair and fingernails, in which it can do no harm. When Napoleon's body was exhumed from its grave in St Helena for reburial in France, it was found to be in an almost uncorrupted condition. Large amounts of arsenic were found in his hair, so much so that it has been suggested that he was deliberately poisoned, although other theories include the idea that the arsenical dyes in the wallpaper of his house in St Helena were responsible for his death. Be that as it may, the arsenic in Napoleon's body was clearly responsible for its state of preservation.

The tests carried out on Napoleon's hair made it clear that the arsenic was not a contaminant from the soil, but was an internal component of the hair. However, the possibility that an exhumed body may have picked up arsenic from the soil around it must always be borne in mind, and there are cases on record in which this has happened.

This chapter has only touched upon its subject, there being so many poisons and potential poisons in existence. Whole groups of poisons, such as abortifacient drugs, bacterial toxins, anabolic steroids and others are worth mentioning, but a full treatment of these is outside the scope of this book, since they are infrequently encountered in forensic case-work.

CHAPTER SEVEN

DESTRUCTION

Man measures his strength by his destructiveness.

Bernard Shaw

I was standing with a police inspector, together with five of his men, in one of the less salubrious parts of London. We stood in a circle around a burnt-out car, which we had arrived at to examine for evidence. The previous day, the car had been undamaged, today it was an ugly, pungent-smelling shell. Whoever was responsible for the crime must have heard about police intentions to examine the vehicle and burnt it as a way of destroying any incriminating evidence.

This is the obvious and fundamental difficulty about the forensic investigation of fire – it destroys even the evidence of its own origin. But not wholly so; fire, destructive though it is, can be investigated, although its investigation must be one of the most difficult in forensic science.

What is fire? I am often asked this question. What is this thing that clearly exists, yet is not an "object"; this thing that is so elusive, much more so than air and water? We see it, we are burned when we touch it, we see the consequences of its action almost every day, when we cook or light the fire, yet we do not know what it is.

To understand the nature of fire we must look at the chemistry of oxidation, an example of which is what happens when oxygen reacts with iron to form rust. No fire is produced by this reaction, since it can happen at low temperatures, but many oxidations need much greater heat in order to take place. In the case of iron, what we call rust is the oxide of iron. The reaction can be expressed in the following way:

$$Fe + O_2 \longrightarrow Fe_2O_3 \text{ (rust)}$$

To balance the equation, we add numbers denoting the proportions of iron and oxygen needed for the reaction to take place, thus:

$$4Fe + 3O_2 \longrightarrow 2Fe_2O_3$$

In this reaction oxygen is said to oxidize the iron. When the gas methane reacts with oxygen, producing carbon dioxide and water, the equation can be expressed as follows:

$$CH_4 + 2O_2 \longrightarrow CO_2 + 2H_2O$$

However, this is not all that happens, for energy, in the form of heat, is also released, so the equation can be more accurately represented like this:

$$CH_4 + 2O_2 \longrightarrow CO_2 + 2H_2O + Heat$$

It can be seen that one part methane and two parts oxygen are necessary for the reaction to proceed in such a way that nothing will be left over from the original chemical components.

Traditionally, oxidation was the addition of oxygen to a chemical substance, while its opposite, so to speak, was reduction, which was the addition of hydrogen. It is now known that such reactions, known as redox (reduction-oxidation) reactions, are characterized by the loss of electrons from one chemical component of a reaction and the gain of electrons by another component. Thus, a substance is said to be reduced when it accepts electrons and oxidized when it loses electrons. Another attribute of redox reactions is that they release more energy than is used up in the reactions themselves.

Nevertheless, all reactions need an input of energy to start them. In the case of the rusting of iron, the required amount of energy is not great and the reaction can proceed at relatively low temperatures. The energy given of is also small but, like other redox reactions, some energy is given off. Reactions that give off energy are called exothermic reactions; those that do not are called endothermic reactions.

How much energy is required to start a redox reaction? The answer depends on the energy barrier of the material concerned. The energy barrier is the least amount of energy required to start a reaction; in the rusting of iron the barrier is very low, while in the burning of petrol it is quite high (in other words, more heat is required to start it). The point at which the energy barrier is reached is known as the ignition temperature. When what we may call a high barrier reaction takes place, so much heat is released that the reaction can sustain itself without any additional energy input. The excess heat and light energy released is what we call fire. The released energy comes from the breakdown of the chemical bonds that originally held the molecules of the reacting components together.

A redox reaction will produce a flame only if the fuel is in a gaseous state and mixed with oxygen. In the case of liquid fuels, the temperature must be sufficiently high to vaporize the liquid. The lowest temperature at which enough vapour is produced to start a reaction with oxygen is termed the flash point, which is much lower than the ignition temperature. In the case of solid fuels, these will burn only when subjected to temperatures high enough to pyrolyse (a term that simply means "break down by fire") into its gaseous components. When we see wood or coal "burning", it is actually the gaseous products of these solid fuels that are reacting with the oxygen in the atmosphere.

Pyrolysis does not always take place when solid fuels burn, however. Although a flame will be produced only when the fuel is in the gaseous state, a solid fuel may burn without a flame if the temperature is not high enough to pyrolyse the solid. This is what we see in glowing embers or cigarette ends. The burning in such cases takes place on the surface of the solid, a phenomenon commonly referred to as smouldering.

A mixture of a fuel gas and oxygen is needed to start a fire, but the proportions of fuel and oxygen in the mixture have to be right. If there is too much fuel (or, to put it another way, too little oxygen), or too much oxygen (too little fuel), the mixture will not burn. There is a fairly narrow range, known as the flammable range, within which the concentration of fuel must lie for combustion to take place.

Many arsonists have come to grief through their lack of understanding of the above principles. In one case a man murdered his wife with an axe. He then poured petrol over her body and all over the room. Like so many arsonists, he seemed to have carried out this activity in a fairly leisurely manner, allowing enough time for petrol fumes to mix with air within the flammable range. When he lit the match, the explosive mixture in the air ignited, blowing out the front of the house and killing him in the process. I have seen this kind of nonchalant fire-raising enacted in television whodunits, in which the culprit carelessly sloshes the fuel around the room, applies a lit match to the drenched furnishings, then quietly retires from the scene. I imagine that what is really used in such films is water, not petrol, otherwise the scene would quickly become too realistic!

The answer to our question "What is fire?" is this: fire is energy, perceived by us as intense heat and light. Fire is made up of electrons moving at high speed as part of the redox reaction, which, as we have seen, is an exchange of electrons. Fire, then, is not an object, but a process – a process of change – but it is a process we can see and feel, which is why it seems so strange and enigmatic to us. Other energy processes, such as the motion of a car, are not perceived by us as being separate from the car itself; we do not perceive or see motion as a separate entity, in the way we perceive fire.

Let us look at one final equation that makes this point clear. When oxygen and hydrogen combine a great amount of energy – fire – is released, as follows:

$$O_2- + H_2 \longrightarrow H_2O + 2e-$$

This shows that when one atom of oxygen reacts with two atoms of hydrogen, two electrons (e- is the symbol for an electron) are released by the oxygen. It is this reaction that caused the explosion and loss of life on the Zeppelin airship, the Hindenburg, in 1937. It was kept aloft by hydrogen, which reacted with the oxygen in the atmosphere, resulting in the disaster in which 36 people died. Hydrogen is now no longer used in airships; helium, an inert gas, is used instead.

WHY FIRES ARE SET

Before we look at the way fire is investigated forensically, it is worth asking why some people deliberately set out fire-raising. The answer, inevitably, is that there are all sorts of reasons, but one of the commonest is the desire to conceal another crime, such as when a murder is committed and the culprit wishes to destroy the evidence, including the body itself or, alternatively, to give the impression that the deceased died as a consequence of what the murderer hopes will be taken as an accidental fire.

An example of attempted concealment of murder occurred when a young prostitute was found dead in her room. There had been a fire and the body was badly burnt. The fire appeared to have started beside the bed, on the right hand side. The impression was that the woman had been smoking in bed and that the fire started when she dropped her cigarette on to the carpet. Fortunately, the woman's four-year-old daughter was not injured and she gave a most interesting account of the affair to the police. She said: "One of the daddies was lighting mummy with matches." Post-mortem examination of the body revealed that the woman had been strangled with a ligature around her neck.

Rarely are fires started in order to *commit* murder – it is too hit-and-miss a method for that. Suicide is rarely committed by fire, since it is clearly too painful a way to die. However, some people have committed suicide in this way, most notably political protesters, who set themselves alight in public.

More commonly, however, arson is committed for economic reasons. A robber might burn down a building that had contained stolen goods; the fire might then be seen as the cause of the disappearance of those goods and robbery may not be suspected, or so the robber's reasoning may go. The records or actual merchandise of a business may be destroyed by a dishonest employee, who hopes to cover up his fraudulent activities. Someone wishing to cheat an insurance company may burn his own belongings.

Evidence that arson is most commonly committed for economic reasons comes from the fact that the incidence of suspicious fires rises during times of economic difficulty. This happened in the USA during the Great Depression of the 1930s; and, in Britain, the recession of the early 1990s saw an increase in suspicious fires, estimated as being a quarter of all fires that happened during that period.

Malice and revenge are other motives for arson. A sacked employee, an insulted or humiliated former friend or a jilted lover may feel sufficiently resentful to exact retribution by means of the destruction of the offending person's property. The burning desire to see justice done may have more literal consequences.

There are other, more complex, motives for arson. Fire has a fascination all its own; most people like to watch a fire burn, even if it is only the fire burning in the hearth. It is a feeling akin to the fascination of torrential rain, thunder and lightning, and stormy seas. These things arouse primaeval feelings in most people and no harm comes from them. It is when such feelings are perverted into a desire to cause havoc quite deliberately that such instincts can lead to crime. The desire to destroy for no actual gain is a much commoner reason for arson, the destruction of tombstones, the defacing of monuments and other acts of sheer vandalism than most people suppose. Most forest fires are caused by human activity, although many of these are caused by accident. Why pyromaniacs and others bent on destruction exist at all is a question we shall leave to the last chapter.

Occasionally, a more novel motive for arson manifests itself. One of the most dramatic examples occurred in England in 1930 when Alfred Arthur Rouse, a former soldier who fought in the First World War, heard of the murder of a woman named Agnes Kesson. She had been strangled to death by a rope, but her murderer was never found.

Rouse was what used to be called a cad, and his caddish activities adversely affected his pocket. He had children in several places, including one in France, dating from the war years. The financial upkeep of these children had to be shouldered by Rouse, and there were two maintenance orders out against him. He was in a social and financial mess and was on the brink of disaster when he tried to deceive yet another young woman, Ivy Jenkins, who was far more intelligent and able than most of Rouse's women friends, who were usually friendless and vulnerable. Ivy's father and brother were quite capable of dealing firmly with her deceiver.

It was then that Rouse thought of Agnes Kesson and how someone had got away with murder. It gave him the idea and he conceived a brilliant plan – a plan of substitution by fire. It was time for him to disappear and rise again in another identity.

In the early hours of November 6, when the last revellers of Guy Fawkes night were returning home, two young men (they were cousins) saw a curious thing in the village of Hardingstone, near Northampton. It was a car burning; and there was someone sitting in it, clearly dead.

It so happens that the father of one of the young men was the local constable, Hedley Bailey. Bailey was roused from bed just after 2.00 am by his son, who told him the news; and both hurried to the scene, where they found that young Bailey's cousin, Alfred Brown, had brought another constable to see the burning car.

At that early stage the police officers did not suspect foul play, but the story the two cousins had to tell raised their suspicions, for they had seen clearly a man acting in a suspicious manner. He had walked past the two men, who had asked him what was happening. He called over his shoulder: "It looks as if somebody has got a bonfire up there."

The fire had not touched the registration plates and the police traced the car and its owner very easily. Rouse was eventually arrested and the two cousins identified him as the man they had seen walking away from the blaze on the night in question. Rouse was accused of murder and put on trial.

In his confession, published after his death, Rouse said he had murdered an innocent victim in the hope that his fire-scorched body would be mistaken for his own. He himself would disappear under an assumed name and would start a new life free of financial and social entanglements. His own story, initially at least, was that he had given a man a lift and, when he (Rouse) needed to leave the car to relieve himself, he asked the man to add more petrol to the car from a can. The man had been smoking cigarettes and Rouse claimed that he must have lit up while holding the can inside the car, starting the fire.

The point that the prosecution had to prove was that Rouse had incapacitated the man in some fashion before he set fire to the car. The only clue to this was the fact that the man lay slumped across the front seats of the car, face-down. This is not a natural posture for someone who was not incapacitated and the prosecution argued that this demonstrated that Rouse had in some fashion prevented the man from escaping. The medical evidence suggested that death was due to shock and burns, which in turn suggested that the man had been alive when the fire started, a view that supported Rouse's own (original) account. In the end, it was not possible to say whether the deceased had been alive or dead when the fire began. The jury found Rouse guilty and he was sentenced to death and hanged. His victim, however, was never identified. According to Rouse's confession, the man was a drifter who "would not be missed". In this horrible prediction Rouse was right. Nobody ever claimed to know the dead man – even Rouse himself gave him no name.

HAZARDS OF INVESTIGATING ARSON

It is not only the destructive power of fire that makes arson such a difficult crime to investigate, for the investigators themselves are subjected to grave risks. A burnt-out building is in constant danger of collapsing and anyone working at the scene could easily be injured or killed by falling beams or walls. Consequently, the first priority is to secure the building and render it safe, before any investigation into the causes of the fire is made. The methods used to achieve this aim may, of course, result in further destruction of evidence, but it is, clearly, necessary.

Some of the damage may not be outwardly visible. Broken or leaning walls or floorboards are easy to see and, probably, to avoid, but walls may crack internally and collapse suddenly for no apparent reason. The reason this happens is that the fire will have heated the fabric of the walls; after the fire the walls will cool down, generating stresses within the stone- or brick-work, which may then crack, causing a wall to collapse. The fact that ceilings and walls are very likely to be unsupported may not be immediately obvious, but any agitation, such as someone walking over a floor, may result in sudden disaster.

There are other hazards, such as the presence of broken glass, nails, sharp edges of stones and uneven surfaces, or surfaces that may appear solid enough, but which, when stood upon, collapse to expose hot, smouldering material into which the unwary investigator can easily fall.

For these reasons, a structural engineer is often asked to assess the nature of the damage and to make recommendations as to how the building should be secured before any investigative works begins. However, there are often more hazards still. Asbestos insulation may have been used in the building and have been exposed during the fire. Noxious fumes and gases, such as carbon dioxide (*see Chapter 6: Poison*) may be present. Measures will have to be taken to ensure that the forensic and police teams are not exposed to the risks posed by such harmful substances. Special protective clothing and masks may have to be worn.

In short, it usually takes some time before a fire can be investigated. In a metaphorical as well as a literal sense, the evidence, never superabundant to begin with, may have gone cold. Nevertheless, some information may be gleaned, even before the investigators enter the building and even when the fire is still burning. It is not often that arson specialists arrive so early at the scene, but photographs or video recordings taken by others may give some indication of where the fire started – the so-called "seat" of the fire. Observations made of the colour of the fire may sometimes be of use, since some substances burn with distinctively-coloured flames. For example, acetone produces a blue flame. Also, the colour of the smoke differs. It is pale grey when benzene burns, but it is black when it comes from burning rubber. These indications are not fail-safe diagnostic methods, but they may suggest possibilities that can later be investigated.

EXAMINING THE SCENE OF A FIRE

In an earlier chapter, I described the "scene" as being, usually, a mess. No scene is a greater mess than the site of a fire, as may be imagined. I am sometimes asked how a fire can be investigated at all. It is difficult, and I have to say at the outset and at the risk of offending many fire-investigators that forensic reports on fires have sometimes struck me as being exceedingly ill-advised, not to say harmful. By this, I do not mean to say that they are fraudulent (although charlatans exist in every profession), but that, in the absence of any other evidence, too much reliance is placed upon them by courts and unwarranted conclusions can be drawn. The available evidence is so scanty, the conclusions so confident, the potential consequences so catastrophic, that it is wise to view the evidence with extreme caution. In fact, experienced fire specialists couch their findings in very guarded terms indeed, but, unfortunately, this does not always happen.

The first thing that must be established during an investigation is where the fire started. When a fire starts to burn in a building, the structural wooden beams will be subjected to high temperatures for a longer time than will such woodwork in other parts of the building, so the extent of charring will be greater in the former than in the latter. Speaking very generally, the lowest point in which the results of intense burning can be observed is probably the point of origin, since fire tends to spread upward. This same characteristic, however, may also complicate the issue, because wood at a higher level may exhibit more charring as a result of the upward progression of fire. It must also be borne in mind that falling, burning woodwork may ignite lower levels where the fire may not have started, although the extent of charring may reveal that the fire did not start at that level.

Other complicating factors may be at work. For example, draughts and wind currents may cause the fire to spread in a way that differs from the norm; the very architecture of a building may encourage the fire to take a certain unpredictable course, and so forth. Experienced investigators are able to note these points and interpret the evidence accordingly.

The intensity of the fire at the point of origin can also be gauged by the extent of damage to other components of the building. For example, glass will distort when subjected to temperatures of over 700°C, while at temperatures above 850°C it will flow and trickle; in this state it has been described as resembling frozen treacle. The steel beams used in modern buildings may also distort at high temperatures and very severe distortion will usually occur at the seat of the fire.

It is sometimes possible to detect the direction in which the fire moved by examining certain objects that have suffered greater fire damage on one side than the other. Plaster may break away from the brickwork when subjected to high temperatures and such spalling may give an indication of the intensity of the fire at particular places within the building. Smoke detectors, if present, may record the order in which the detectors were set off and may be recovered and the information taken from them noted.

All sorts of little clues can help to reconstruct the events of a fire. Smoke blackening at the top of a door indicates that the door was open during the fire, otherwise blackening would have occurred over the whole surface. Thin lines of severe burning on the ground, surrounded by areas of less severe burning, indicate that a trail of gasoline was laid by an arsonist. A large mound of combustible materials in one place may also suggest arson, as would the absence of valuables (jewels, documents, money, etc.) from their usual places. Signs of forcible entry, such as evidence of doors forced open; broken windows with glass fragments lying inside the building; broken locks and so forth, will also point to deliberate fire-raising. The number of such details is almost endless and the more experienced the fire investigator is, the more such clues are likely to be noticed.

The search for faulty electrical equipment, which may have been the cause of an accidental fire, must also be a priority. In spite of the destructive nature of fire, it is often possible to find clues in electrical appliances, such as whether regulators were switched on or off. Fuses can also be examined in order to determine whether they had failed. Similarly, gas appliances may cause fires if they had been misused, such as when clothes are placed upon them for drying.

From this general review of the kinds of evidence sought by forensic scientists, it can be seen that the investigation of fire is more of an art than a science. To say this is not to demean the subject – quite the contrary – but it does mean that to become an effective fire investigator is not a simple matter of training, but it is a skill that is acquired through years of experience. Hard science certainly comes into it, but in more subtle ways than it does in other areas of forensic work.

Apart from such things as signs of forced entry and the absence of jewellery, what arouses suspicion of foul play in the mind of a fire investigator? The answer, of course, is evidence that the fire was caused deliberately, but what constitutes such evidence? First, it may be concluded that there was more than one origin of the fire. An accidental fire will have only one source – at least, it is extremely unlikely for two accidental fires to start at the same time in the same building – so when the evidence appears to show that the fire began in more than one spot, then it is justified to believe that the fire had been deliberately started. For example, if one should find that two widely separated places within a building show signs of long-term burning and other evidence of the kind discussed above, then it is likely that one is dealing with a case of arson.

THE USE OF ACCELERANTS

Arsonists hardly ever simply strike a match to light a fire, using any combustible material to hand, such as a piece of paper or a curtain. Such a course of action is too uncertain, since a fire lit in this way may burn itself out very quickly. Usually, an accelerant is used. A flammable liquid such as kerosene is poured over a wide area of carpets and furnishings, before the match is applied. This ensures that a hot fire will follow and that the building will be ablaze long before any fire-fighters arrive. However, what most arsonists do not know is that traces of such accelerants can be detected, even after the fire has destroyed the building. Small amounts of accelerant will seep into carpets, floorboards, plaster, brickwork and other materials and will not be consumed by the fire. The cooling effect of the water used to quench the fire will slow down the rate of evaporation of the accelerant and enough will usually remain to be detected.

"Sniffing" devices – essentially, hydrocarbon detectors – are often used to test objects at the scene to see whether they contain accelerants. A long nozzle is attached via a cable to a box with a meter. Air around the object in question – say, a piece of floor-boarding – is sucked into the nozzle and into the box, where it passes over a heated filament. If an accelerant is present, it will be oxidized and will raise the temperature of the filament, the increase in temperature being recorded as a deflection of the pointer on the meter. Such a device is used as a preliminary screening method, to show whether more detailed laboratory analysis is worthwhile.

Items, such as pieces of carpet, sections of floorboards or fragments of upholstery, which may contain accelerants and which might benefit from further laboratory examination, are packaged in airtight containers. Small items can be kept in glass jars with tightly-fitting lids, but larger items should not be carried in polythene bags, as these will react with any hydrocarbons present in the item. Usually, large items are broken into smaller pieces so that they can be placed in jars. however, if this is not practicable, specially made bags may be used.

Apart from "suspect" items of evidence, the collection of similar, but unburnt, objects from other parts of the fire scene is useful. This serves several purposes: first, these objects may reveal the presence of larger amounts of accelerants, making laboratory analysis easier. On the other hand, tests on such samples may show a complete absence of accelerants. Such a finding would be useful evidentially, since it could otherwise be argued that the presence of flammable substances was due to the innocent presence of, say, carpet cleaning fluids.

Control samples have yet another function. Intense heat may result in the breakdown of linoleum, glues, plastics and similar materials, releasing hydrocarbons of a kind that could be mistaken for accelerants. It is clear that the interpretation of results from hydrocarbon analysis must be made with great caution.

The clothing of arson suspects must also be examined for accelerants, since arsonists hardly ever fail to spill some of the fluid on to themselves. However, the sensitivity of some of the laboratory techniques is such that innocent people may be incriminated, because sweat may react with certain fabrics, producing volatile substances similar to the ones in commonly used accelerants. Clothing can also provide clues of another kind, since synthetic fibres can melt in a distinctive manner in response to heat. Such melting may not always be obvious to the naked eye, but microscopic examination can reveal that it has taken place. Again, it is important to be aware that such damage can have innocent causes, since not only can innocent burning or melting take place, but also the effects of certain fabric cleaners and other chemicals can produce changes that may be confused with heat damage.

Once material suspected of harbouring accelerants arrives in the laboratory, the items are subjected to techniques aimed at isolating the hydrocarbons. The item to be examined is placed in an airtight jar and heated, releasing any volatile hydrocarbons, which are removed with a syringe attached to the apparatus, then introduced into a gas chromatogram, the resulting chromatogram allowing the identification of the chemicals (see Chapter 6: Poison). Another isolation technique is to heat the suspect item in the presence of a charcoal-coated Teflon strip, on to which the hydrocarbons will adsorb. This method of concentrating the vapour allows the isolation of a much greater amount of hydrocarbons. The adsorbent strip is then washed in carbon disulphide, which will dissolve the volatiles. The mixture is then injected into the gas chromatogram for identification. The printout for carbon disulphide is known and can be disregarded, the investigator considering only the additional peaks that appear in the chromatograph.

The commonest accelerants used by arsonists are petrol, paraffin, methylated spirits, alcohol, turpentine, diesel and other easily available organic solvents. However, the investigator must always be aware that the unexpected may arise.

If an incendiary device had been used in the arson attack, its remains may be found at the scene. The wick and shattered bottle of a Molotov cocktail, for example, can often be found. Even a discarded match may survive the fire unburned, if it is cast aside some distance from the fire. Diatoms (small, single-celled marine algae with silica shells) are found in the so-called diatomaceous earth deposits, which are used in the manufacture of match-heads. The silica shells are abrasive and help with the striking of the match. Different brands of matches contain different mixtures of diatom species, which can be identified from their microscopic shells. Thus, the brand of match can also be identified. In the USA, match "books" are more commonly used than in Britain, and a discarded match torn from such a book can sometimes be matched (no pun intended!) with the book from which it came.

As so often happens in forensic work, there arises a different kind of evidence rarely encountered in a particular specialism. An example took place when the premises of the Branch Davidians at Mount Carmel, near Waco, Texas, were assaulted by FBI agents on April 19, 1993. The ensuing fire killed seventy-six members, including women and children.*

The dispute that ended with the assault was concerned with the fact that the Branch Davidians possessed large amounts of firearms and explosives illegally. The attack, consisting of shooting CS "ferret" rounds, as well as the spraying of tear gas, began just before 6.00 am. The FBI said that the firing stopped at 11.40 am; and that at 12.07 pm three or four fires started suddenly. The building was destroyed within a few minutes.

Later, the surviving Branch Davidians claimed that the shooting continued as the building was burning down. An FBI aeroplane film-recording showed flashes of light emanating from the FBI positions, but it was unclear what caused them. An experiment, carried out in 2000 by a British company, Vector Data Systems, simulated what may have happened. Eight shooters, using several different kinds of weapon, were filmed from two circling aeroplanes equipped with special heat-sensing cameras. The images were then compared with those obtained during the attack itself.

The results were said to show that the flashes in the original images lasted much longer than those that were caused by the firing of weapons in the experiment. It was concluded that reflected sunlight from debris was the most likely cause of the flashes and that there was no evidence that gunfire continued during the fire itself. It ought to be said that this result was disputed by the Branch Davidians' lawyers, and the situation has continued to involve controversy over the part played by the Law Enforcement Agencies.

SPONTANEOUS COMBUSTION

Horrific conflagrations resulting in many deaths always make the headlines. This is perfectly natural, but such terrible events seems to be associated with a certain kind of mythology unique to them. The Cambridge botanist Oliver Rackham coined a very useful word; he referred to "factoids" in science. A factoid is a statement that is so familiar and so often repeated that no one doubts it. It is also totally false. One such factoid is the notion of spontaneous human combustion.

When the story of a mass disaster in the form of a terrible conflagration reaches the newspaper headlines, it is often stated that the surrounding temperatures were so high that some individuals spontaneously ignited and were burnt, even though they did not come into contact with the flames. On other occasions, when a human body is found displaying severe local burning, but where there is little or no fire damage in the surrounding area, "spontaneous" human combustion is very frequently invoked, the "cause" being a supernatural "internal" source of fire. Spontaneous human combustion is sometimes the explanation offered by an arsonist; such a theory from such a person, while not excusable, is certainly understandable. What is almost impossible to understand is the fact that some scientists and very many journalists, as well as most of the general population, seem wedded to the idea of spontaneous combustion, despite the rebuttals issued by forensic scientists whenever a horrific fire makes the headlines.

There is no scientific evidence whatsoever for spontaneous human combustion. It is a complete fiction, but it is still believed in very strongly by so many people, including some scientists. This is not to say that spontaneous combustion – as opposed to spontaneous human combustion – cannot happen; it does happen under certain conditions that are very different from the conditions that obtain in a human body. For example, a large mass of hay or compost may spontaneously ignite at high temperatures. This happens because redox reactions are exothermic, producing more heat than they consume. Also, when a large mass of material burns, the amount of reacting components is very large in relation to the surface area of the mass, unlike the situation in small masses. This means that much of the heat released cannot be dissipated quickly through the surface, which is the only area from which dissipation can take place. The result is that the retained heat raises the temperature, causing the reaction to proceed at an even faster rate, which will then release more heat and so on. The conditions required for spontaneous combustion cannot exist in a living human body.

Spontaneous combustion of materials such as decaying vegetation or oil-soaked rags is sometimes the cause of a fire; and a search for a possible source of such ignition should be made. In cases in which a person is found dead with localized burning, but where the surroundings are untouched by fire, the cause is usually the fact that the deceased had caught fire near a flame, such as a coal-fire or an oven, but then moved away from it and collapsed.

CONSIDERING EYEWITNESS EVIDENCE

There is one particular class of evidence often sought by fire investigators, but it is a kind of evidence which, in my view, is highly undesirable for a scientist to consider. It is eyewitness evidence. In so many cases from my experience, this kind of evidence introduced more confusion into forensic investigations than any other kind of evidence sought by scientists. I believe it is a kind of evidence, generally speaking, that scientists should treat with great caution.

Such a strident assertion requires justification, especially since most academic textbooks on arson investigation devote substantial sections to the consideration of eyewitness evidence. Fire investigators are advised to note what witnesses say and take such statements into consideration when writing their reports. In part, this is due to the fact that evidence is so hard to glean after a fire and investigators feel they have to rely on whatever evidence they can get. However, it is also a consequence of the fact that a fire, unlike most murders or robberies, is such a dramatic event that is so often witnessed by passers-by, that it may seem perverse to ignore the eyewitness evidence that is almost always available.

Why, then, do I say that such evidence should be kept at arm's length? First, I must make it clear that I am here referring to what the forensic scientist, not the police officer, should consider. The latter should, of course, take great interest in such evidence, but, if evidence is evidence, is not what is sauce for the goose, sauce for the gander? No, for the simple reason that the forensic scientist is usually asked for an opinion on which particular witness account is the more likely. To make the point clearer, consider what a scientist, asked for an opinion about the time of death in a murder, should do. Are eyewitness reports to be used when a scientist is considering the time of death on the basis of maggot evidence or the onset of rigor mortis? Clearly, the answer must be no, since the whole point of consulting a forensic scientist is to obtain an independent opinion about the course of events. It would be strange, to say the least, if a pathologist were to arrive at a time of death estimation after studying the rate of drop of temperature of the body, then ask an eyewitness for their account and amend his report accordingly. I think anyone, including all lay people, would think this a most unusual way of conducting a criminal investigation.

Yet this, in essence, is what is considered normal in a fire investigation. I have seen so many fire reports that say something along the lines that, since a certain eyewitness saw someone standing at a window holding a canister of what must surely have contained petrol, then the house must have been set alight at that spot. This kind of information, while interesting to the police officer, is not the kind of impartial evidence expected from a forensic scientist. One does not need a scientist to repeat the testimony of an eyewitness, rather the scientist is required to comment on whether the eyewitness's statement is more likely to be true or false.

Yet have I not said earlier in this book that a forensic investigator should use whatever evidence is available? Yes, indeed, but one must be careful to distinguish what one may call "mute" evidence – facts – from opinion. It is, of course, quite possible, even probable, that an eyewitness may be speaking the truth and so their account would consist of facts, not opinions. However, it is not possible to know. In court, one may have to answer questions of the kind: "If X really did see Y at the window, would that alter your conclusions in any way?". In reply to such a question, one should say that, if X really did see Y, then one can say that such-and-such a thing would have been likely to have happened, but that, as a forensic scientist, one cannot have an opinion on whether what X said was true or not. In short, one cannot use X's statement as any part of the starting point of the scientific investigation. X's observations are a matter for policemen and juries, not for forensic scientists.

Where, then, does this leave such things as documentary and psychological evidence? Are these not, in a way, eyewitness evidence? Do I say that these areas of forensic concern cannot be investigated scientifically? Not at all; but we are dealing here with a subtle point. A document may, indeed, include untruthful statements, but it is the veracity of those very statements that forms part of the investigation. In other words, a lie can be evidence, depending on its context. Here, one is not looking upon the document as a source of information, rather one is assessing the reliability of the document itself. A lie or, at least, an incorrect statement may provide evidence of duplicity, madness, a warped sense of humour, a defective memory or an honest mistake. We will go more deeply into these matters in the next chapter.

THE SIGNIFICANCE OF CARBON MONOXIDE

People caught in a fire often behave in an unpredictable fashion. Sometimes it may appear odd that they were unable to escape, when the scene is examined after the fire had been put out. The smoke, which impairs visibility, as well as the sheer terror of the situation may make a person rush toward the fire rather than away from it. The inhalation of carbon monoxide (CO) will reduce the ability of the victim to flee, since the reduced amount of haemoglobin available for the carrying of oxygen will result in muscular weakening, leading to collapse (*see Chapter 6: Poison*).

CO blood concentrations of about twenty to thirty per cent make the sufferer feel unwell, with headaches and dizziness. Forty per cent saturation will result in a lack of muscular coordination and general mental confusion, with the sufferer staggering about, unable to make properly coordinated movements. These effects become more pronounced at fifty per cent, when slurred speech, haphazard, aimless movements, extreme weakness and vomiting take place. By about sixty per cent the victim will lose consciousness. Death follows rapidly at concentrations of around seventy per cent.

These figures are only rough guides, since responses vary from person to person. On September 28, 1902, the French author Emile Zola died during his sleep as a result of CO poisoning. There had been a fire burning in the grate and it is said that the chimney had not been swept or was faulty in some respect, although others claimed that it was deliberately blocked in order to commit murder. (Zola had many enemies, especially after his defence of Dreyfus.) Interestingly, Zola's wife was found unconscious, but she recovered later.

Absence of CO from the blood in the bodies of people discovered after a fire is very suspicious and points to foul play. For example, the prostitute in the case mentioned earlier had no trace of CO in her blood, which indicated that she was already dead before the fire began. Anyone dying in a fire will have a certain amount of CO concentration in the blood.

Since the Second World War the fabric of most upholstered furnishings has been manufactured from substances which, when burnt, release quantities of cyanide gas (HCN). Clearly, death in such cases takes place very rapidly indeed. It is puzzling that this problem has not been addressed by manufacturers or governments, since one imagines that other, less harmful substances could be used in the production of furnishings.

Another tell-tale sign is the condition of that part of the body (or clothing) that is in contact with the floor. If the clothes or skin in those areas are free from fire damage, the conclusion must be that the victim had been in that position before the fire began, or at least fell to the ground during the very early stages of the fire. Such evidence may be useful in cases in which it is uncertain whether the fire was the actual cause of death, although other evidence, such as blood analysis of CO and HCN, will be have to be done before any definite conclusion can be reached.

A case that brought together many aspects of arson and murder investigation took place in St Lucia during the summer of 1971. On June 10 a house on the mountainside near the port of Castries burst into flames. The dilapidated fire engine took a long time to reach the house and, on arrival, the firemen found the house effectively burnt out.

Two charred bodies were found in the bedroom, and were assumed to be those of the owners, James and Marjorie Etherington. At first, no-one suspected foul play. It was taken for granted that the fire was accidental and that the victims died as a result. The bodies were taken to nearby Barbados for burial.

As often happens, it was an insurance man who suspected arson. In typical Sherlock Holmes fashion, he went sniffing around the house and found soil footprints on the floor of the rear scullery. Moreover, the footprints were beneath a window that had been smashed inward. Nosing around further, the man found a green plastic hose running into the house over the sill of another window. Putting the tip of the hose to his nose, he detected a strong smell of petrol. He then traced the hose back to the garage, where a car stood with its petrol tank open, the cap lying on the floor. All this amounted to a great deal of evidence that had been overlooked both by the police and the firemen!

When the insurance agent made all this known, Scotland Yard and forensic specialists from Britain were called in to investigate the case. The arson specialist was able to identify at least four, possibly five, seats of fire; and evidence of accelerant hydrocarbons were found in the floorboards.

At this point, the police arrested three suspects – men who were routinely hauled in whenever a crime was committed, so notorious were their reputations. Their names were Florius, Faucher and Charles. Florius, the ring-leader, had burn marks on his neck and arm and he and Charles had fingernail scratches on various parts of their bodies. All three men admitted to the robbery, but denied killing the couple or burning the house. They just tied them up, they said, then left. Then, in the classic manner of the cocksure criminal, Florius challenged the police to prove him false.

They did. Apart from all the evidence already amassed, post-mortem examination of the exhumed bodies revealed that Mr Etherington's skull had been smashed with a blunt instrument; and a clot of blood showed that there had been surface bruising of the brain. Soot particles were found inside the lungs, suggesting death was due to CO poisoning. More ominously, the examination of Mrs Etherington's body revealed the presence of a clothes line running between her teeth, gagging her in life. Laboratory analysis of both blood and muscle samples from the bodies confirmed the presence of CO.

At the trial, the defendants gave the game away, for Faucher said that Florius had told him that they should burn the bodies, "as in England they have a new method to make the dead talk" and that "if we burn them, ashes don't talk". Whatever Florius thought he meant by all this is not clear, but the remains did talk "very eloquently", in the words of the Attorney General, Mr John Renwick. All three men were found guilty and hanged.

EXPLOSIVES

We have seen that mixtures of air and petrol, if in the correct proportions, can become explosive when lit. We will now look at substances that are designed to be explosive.

There are two kinds of explosive: low explosives and high explosives. Low explosives are characterized by their ability to burn like any other combustible material under normal circumstances; they only become explosive when they are confined in a small space. The archetypal low explosives are black powder (commonly called gunpowder) and smokeless powder. All low explosives are essentially mixtures of an oxidizing agent and a fuel.

Black powder is a mixture of saltpetre (potassium nitrate), charcoal and sulphur. It can be used as a fuse to ignite a larger amount of powder confined in a container. Smokeless powder is made of nitrocellulose (cotton treated with nitric acid – "nitrated cotton"), or a mixture of nitroglycerine and nitrocellulose. It is a more powerful explosive than black powder.

When a low explosive explodes, it exerts a throwing effect – objects are hurled about, walls are blown asunder and so forth. Mixtures of air and gaseous fuel explode with a similar effect, but they can have an additional burning effect if the flammable mixture is at the high end of the range. In such cases the mixture will explode, but some gas will remain unconsumed and, as air rushes back to the area of the explosion, oxygen will combine with the remaining hot gas and a fire will start to burn. Often, it is this secondary fire that causes most of the damage, not the explosion itself.

It is the sudden generation and expansion of gases within a container that causes the explosion. The gases can rush out at speeds of more than 7,000 miles per hour, which are forces sufficient to make buildings collapse and to throw large objects, like boulders or cars, into the air.

High explosives are far more damaging. They explode at rates of between 1,000-8,500 metres per second and their effect is literally a shattering, rather than a throwing, one. In other words, a car exposed to a high explosive will be smashed to pieces, rather than lifted and thrown. High explosives are, in fact, of two kinds, both of which are used in a high explosive system. The first group consists of extremely heat-, shock- or friction-sensitive compounds, such as mercury fulminate or lead styphnate. They detonate violently, even when not contained in a confined space. In view of their extreme sensitivity, they are not used in the main charge, but are used to initiate the explosion. In other words, they are detonators, or primers. Most modern detonators are set off by an electric current that passes through a wire filament that ignites the fuse-head, which in turn ignites the priming charge. This latter then ignites the base-charge of secondary explosive, which amplifies the effect, causing the main explosive charge to detonate.

The main charge of a high explosive is usually a substance that burns, rather than explodes, if lit in small quantities in the presence of air. The most famous of these non-initiating explosives include dynamite (nitroglycerine mixed with an absorbent material, such as kieselguhr), TNT (trinitrotoluene), PETN (pentaerythritol tetranitrate) and RDX (cyclotrimethylenetrinitramine). TATP (triacetone triperoxide) is a 'home-made' explosive that can be used both as a detonator and as a main charge. It is often used by terrorists in improvised explosive devices (IEDs). A TATP-based bomb was used in the London bus bombings of July 2005.

When fragments of explosive are collected from a scene, the pieces are washed in acetone, which dissolves most of the chemical components. Having isolated the compounds in this way, chromatographic techniques are used to identify the individual compounds. Once this is done, knowledge of the composition of various explosives allows the forensic specialist to identify the particular make of explosive used.

THE LOCKERBIE DISASTER

As with so much forensic work, the best way to describe how an explosion is investigated is to look at a specific case. We will take as our example the worst disaster in British aviation history.

A little after seven o'clock on the evening of December 21, 1988, the Boeing airliner 747-121 (Pan Am Flight 103), on its way to New York from London at a height of 31,000 feet, suddenly disintegrated above Lockerbie in Scotland. All 259 passengers, as well as the crew and eleven people on the ground, were killed. The wreckage spread over an area of more than one thousand square miles.

Although the most likely cause of the disaster was a terrorist's bomb, it must be made clear at the outset that this has not been proved. As we shall see, at least one other explanation is possible. I say this, not because I have any real doubts that terrorist activity is to blame, but because I believe it to be a desirable corrective. Subtle, yet very strong, pressure can be brought to bear on forensic investigators involved in a tragedy that has international political implications, and the assumption from the outset was that the destruction of the airliner was an act of terrorism. Even, I should say especially, in highly charged situations like these, pure objectivity is essential.

The first stage of the investigation was the search for the debris of the plane. This was by no means easy since not only was the wreckage spread over a very large area, much of that area was woodland. This hampered the thousand searchers on the ground and also presented difficulties for the large military helicopters, which could not manoeuvre easily above the woods. Later, smaller helicopters were used. They carried infra-red camera equipment, which could detect debris beneath the canopy.

Satellites were also used to photograph the area. The first such photographs were supplied by a French satellite, which, although helpful, were not of a sufficiently high resolution to identify objects on the ground. Computer enhancement methods were used to gain higher resolution, but, in the end, NASA spy satellites photographed the area, with excellent results. It may seem astonishing to say, but these spy satellites can be used to read the text of this book from several miles up in space.

This combined operation – ground searchers, aerial photography and satellite images – resulted in the collection of more than ten thousand objects and pieces of debris. Each item was X-rayed and tested for any residues from explosives. All relevant details from every item were fed into the Home Office's computer, known, delightfully, as HOLMES (Home

Office Large Major Enquiry System). All items that were pieces of the fabric of the aeroplane itself were deposited in a hangar, where the airliner was painstakingly reconstructed. The purpose of this reconstruction was to see how the aeroplane broke apart in the first place, since explosive damage would be most pronounced in those parts that were closest to the bomb, if one had been present.

When a bomb explodes on board an aircraft, the metal in the fuselage will melt. The hot gases generated will punch small holes in the metal. Moreover, the very low temperatures outside an aeroplane travelling at high altitude will lower the temperature of the metal quite suddenly, resulting in a distinctive microscopic metallographic structure. These physical changes to the metal allowed the investigators to narrow down the number of possible sites of the bomb.

The distribution of fragments of the bomb was also a useful clue. Luggage debris was found on the right-side engine of the craft. This, taken together with the assumption that the aeroplane must have broken at its weakest point, which was the cargo door on the right (there was no such door on the left side), allowed the investigators to conclude that the bomb was situated in the left side of the plane, since the debris would have been sucked toward the right-side engine. This argument seems quite reasonable as far as an explanation of the effects of a left-side bomb goes, but it seems to me that it does not exclude the possibility that the bomb had been lying in the right-hand side of the plane, which, in fact, was the original conclusion. In the end, it was decided that the bomb was placed in the left side of the forward cargo hold, just under the "P" of the Pan Am logo. The fact that the cockpit fell to earth in one piece, separated from the rest of the plane, reinforced this belief.

The gas chromatographic analysis of chemicals isolated from the various items showed that the Czech-made explosive Semtex was probably used in the bomb. Some items of clothing had appreciable quantities of compounds that are found in Semtex. However, as we have seen in the context of poisons, many compounds often found in explosives may have an innocent explanation.

A particularly interesting discovery was the finding of small pieces of a stereo system of a make called Toshiba "Bombeat". This make is available only in the Middle East and North Africa, including Libya, upon which country suspicion was soon to fall. Another Libyan link was a microchip from the detonator circuit. This was of a kind identical to ones found on two Libyan agents in Senegal in 1986. The agents had also been carrying twenty pounds of Semtex. Interestingly, many bags from Malta were found in the hold – Malta is a country with strong political ties with Libya.

Fragments of detonators were also found and it became clear that a two-step detonator had been used. The first step was a barometer-detonator, which is triggered by the drop in temperature when the aeroplane reaches high altitudes. The second was an ordinary timer-detonator. The presence of both detonators was probably the reason why the airport authorities failed to detect the presence of the bomb, since terrorists normally use a barometer-detonator only. When a suspicious item is tested at an airport, it is subjected to low pressure, which would detonate the bomb under controlled conditions. If the bomb on board Flight 103 had been treated this way, it would not have gone off.

Many people have asked why the pilot did not send a distress signal, which, normally, he would have had time to do. The answer to this is, if the bomb had, indeed, been placed in the forward cargo hold, then the explosion would have damaged the electronics centre of the plane, rendering any attempt to send radio signals futile. The cockpit voice recorder registered a loud bang less than a second before the aircraft disintegrated. It is quite possible that the sound was that of a bomb exploding.

These are the most important basic facts of the investigation, so why do I say that other causes of the explosion are possible? First, let me say that some proposed causes are so unlikely as to border on the fanciful. One such is the idea that magnetic disturbances were responsible. We need not go into these ideas any further, but it must be realized that aeroplanes sometimes do disintegrate as a consequence of bad weather, bad piloting or by collision with birds, which are sometimes sucked into the engines.

The most realistic alternative explanation for the Lockerbie disaster was proposed by John Barry Smith, a specialist in the investigation of disasters of this kind. He postulated that the cargo door was defective – a defect known to exist in other aircraft and which has caused air crashes. Also, other Boeing 747s have had defective doors. Consequently, Smith believes that there was no bomb on board and that the simplest explanation is the defectiveness of the door. We have already seen that the investigators recognized the door as being the weakest part of that part of the plane, although this would remain true, even if the door had been sound.

How does this fit in with the rest of the evidence – the microchips, the Semtex compounds, the bang on the voice recorder, the Toshiba "Bombeat" stereo, the Maltese bags and the detonators? The answer is that a certain microchip might mean that a person from the Middle East was involved in some way, but it does not mean that that person was specifically a Libyan – in any case to be a Libyan does not mean that one is a terrorist. Objects can be available for sale only in a few countries, but can be given as gifts to others. This applies not only to the microchips, but to the stereo system as well. The compounds (RDX and PETN) found on the various items are not only found in Semtex, but in other explosives and could have been there as a result of contamination. Maltese bags are, in themselves, no indications of guilt. The bang on the voice recorder could have been the sound of the aircraft breaking, rather than the sound of a bomb.

The detonators are harder to explain away. Also, the cumulative nature of the evidence and the fact that the bomb scenario is much more inherently probable than the others compels one to believe that the tragedy was an inhuman terrorist act. Personally, I believe that it was.**

---

*The acrimonious debate about who was to blame for the fire and deaths continues. The evidence appears to me still to be unclear, so I will not discuss the fire as a whole, restricting myself solely to a discussion of one investigative technique used.

** On August 16, 2003, Libya formally admitted responsibility for the Lockerbie bombing and paid compensation to the victims' families. In February 2011, during the civil uprising in Libya, the former Justice Minister Mustafa Abdel-Jalil claimed that Colonel Gaddafi had ordered the bombing and in December 2011, permission was granted by the new Libyan government for further investigations by British police officers.

CHAPTER EIGHT

WORDS AND IMAGES

If you give me six lines written by the hand of the most honest of men, I will find something in them that will hang him.

Cardinal Richelieu

I was sitting at my desk, examining a diary, turning the pages slowly and marvelling. The detective inspector, seated opposite, had a broad smile on his face and was metaphorically rubbing his hands with glee, as well he might. For the document before us belonged to an international criminal, who was wanted in a dozen countries for crimes of robbery, drug-trafficking and murder. And there before us was his diary, recovered from his body after he had been killed by one of his rivals, all of whom were wanted murderers and drug dealers, too.

Names, addresses, telephone numbers – all were in the diary. The diary itself revealed the place where the criminal had bought it. The information was of a kind that would normally have taken a detective months or years to find, if, indeed, he could find it at all. Finding the diary was a rare event in a police officer's life.

There is something particularly awesome about documentary evidence. A few marks on a piece of paper can incriminate or exonerate, bring great happiness or great misery, inform us or reveal to us the extent of our ignorance. Writing is so much taken for granted that we tend to forget the power of it. It is no wonder that, in days gone by, those who could read and write were looked upon as wizards.

In this chapter we will consider what the written word or the image can tell us about a crime or a past event. We will also consider the spoken word, which, when taken in context, can reveal so much about a criminal's thoughts.

HANDWRITING ANALYSIS

A person's handwriting is a very personal attribute and no two people share exactly the same style. This uniqueness is the basis of handwriting identification. However, this does not mean that the differences between the handwriting of different people are always obvious and clear-cut, since many of the differences are very small and subtle. Indeed, it would be surprising if they were not, since there is a limited number of ways in which a letter can be written and each person will adopt one of a small number of styles. The distinguishing features lie in the little things.

There are variations, too, between the handwriting of a single person at different times. Changes in style take place with age – a child's handwriting is very different from that of an adult. As a child grows up, it will change its style considerably, but, even when that style has stabilized in adulthood, changes will still occur. Tiredness, illness, emotional state, position of writing, the drinking of alcohol, among other things, will cause changes in handwriting. Nevertheless, the basic style usually remains detectable.

In forensic cases, the document examiner is usually asked whether a hand-written document was written by a particular suspected individual or not. Comparisons with other undisputed examples of that person's handwriting will have to be made, but what exactly is being compared and how is the comparison done?

First of all, the gross differences are examined. Some people write the fifth letter of the alphabet thus: e. Others may write it like this: – the so-called Greek "e". Some may have a habit of curving the tail of the "y" at the end of a word, thus: y, but others may leave it straight – y. Many other differences of this sort can be imagined.

These differences, although numerous, are finite and cannot be used to establish authorship, although they serve as a rough sorting exercise. If the actual writer of the document wrote Greek "e"s and curved the tails of his "y"s, while the suspected author of the document does not, then the latter can be excluded. On the other hand, if the "e"s in the document were all non-Greek and the "y"s all have straight tails, then this cannot be regarded as indicating authorship, simply because so many people write those letters in that way. Further analysis is required.

We can all recognize the form of someone's handwriting, but we would find it difficult to describe it in a fashion that would enable anyone to recognize it on sight, without having previously seen a sample of it. The document examiner tries to prepare a kind of visual description of the two sets of handwriting being compared. This can be done by comparing the relative positions of the highest points of the letters and the lowest points of the letters, giving us a picture of what we might term the "amplitude" of the writing. This point can be made clear by examining the following two forms of the letter W:

W and W

The first form has the three peaks of the letter at the same level, while the second letter has the middle peak at a lower level than the ones on either side. If we place a dot at the top of each peak, we will get a pattern ... for the first letter and a pattern ... for the second. If the second letter of the word is an h, we can do the same thing to it in relation to the three dots of the W, thus:

Wh and Wh

We will now have the patterns ... and .... respectively. If we continued in this fashion throughout the word, we may find that the patterns of the two scripts appear in like this:

.... and ..

If we subject many other words and sentences to this kind of analysis, we may find a consistent pattern emerging. If, in addition, we do the same thing to the lower ends of the letters, especially letters like y and j, which extend below the imaginary line of writing, we can add further patterns for examination.

We can do more than this. If we draw a line through each stroke of the letter in order to simulate its angle of incline, we will have another kind of pattern for examination. Thus, the two letters we are comparing may appear in two different forms, like this:

W and W

Lines drawn through the strokes of the W will appear something like this:

| | | and | / | /

Similarly, one can place dots at the end of each word and at the beginning of the next one. This will give an idea of the width of the spacings between words – some people write their words quite close together, while others leave wider spaces. Thus, one may find that two different sentences are spaced as follows:

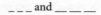

_ _ _ and _ _ _ _

Analysis of this sort is usually done on tracing paper laid over a photocopy, so that the original document is not damaged. In order to see if there is a reasonable match between samples of handwriting, transparencies can be laid over one another and examined. One of the difficulties with identifying handwriting is that there is sometimes no undisputed sample with which it can be compared. In such cases the suspected author of the document is asked to write a piece of text dictated by the document examiner. The danger inherent in this is, of course, that the writer may deliberately alter his writing in order to deceive the investigators. In spite of this, however, certain basic aspects of handwriting style cannot be changed easily, since these are done automatically and the writer may not be aware of these subtleties.

At this point I must draw attention to a serious weakness in the methodology of handwriting specialists. Almost always the opinion given by them is based on a comparison of the suspect document with a known sample from a suspected individual. This is not a scientifically valid practice, since the possibility that there may be another author (i.e. other than the suspect) is not addressed. So, if the handwritings do not match, then the suspect is cleared, but if, in the opinion of the examiner, they do match, then the suspect is guilty. However, it is very difficult to say what, in fact, constitutes a match.

In general, forensic document examiners use a five-point scale in their conclusions about the possibility of whether a particular individual was the author of a document or not. It runs as follows:

1. Common authorship. One hundred per cent positive, written by the same person.
2. High probability. Very strong positive evidence; very unlikely to have been written by a different person.
3. Probably, or could well have been written by the same person.
4. Inconclusive.
5. No evidence. (This does not mean that the suspect did not write the document, merely that there is no evidence that they did.)

The subjectivity of this scale is obvious. The word "probably" means different things to different people. It is therefore not a scale, but a list of subjective opinions that cannot be quantified. It is, indeed, very difficult to quantify similarity of handwriting, but, until such a system can be devised, the above scale must be seen as a method of classifying opinions, not of classifying similarities. To my mind, this, together with the lack of consideration of other possible authors of a document, weakens handwriting evidence considerably.

It may not be obvious why other possible authors should be considered, so let me present the argument in everyday terms. Let us say I give you an apple, then take you to a tree and ask you whether you thought the apple came from that tree. Upon inspecting the tree and finding that it bears oranges, not apples, your answer will be that the apple could not have come from that tree.

If I had taken you to an apple tree that bore yellow apples, you would, upon examining the apple I gave you and finding that it was red, conclude that it did not come from that tree either. But what if I had taken you to an apple tree bearing red apples, like the one I gave you?

I suspect that your answer would have been that the apple could quite easily have come from that tree, but that you cannot be sure, since there are many red-apple trees in existence. You might compare the apple in your hand to the apples on the tree and you would not be able to see any fundamental difference between them.

Of course, the whole exercise was flawed, since I did not give you sufficient information upon which to make a decision. The proper course of action would have been for me to show you as many red-apple trees as possible and ask you which one of them the apple in your hand came from. Although you could find no fundamental differences between your apple and the ones on the single red-apple tree I showed you, you might well have found many small differences between the apples on the many different red-apple trees. This is because any description is incomplete; we describe only what we think is essential or relevant. And what we think is essential or relevant depends on what we are comparing the object with.

I maintain that samples of handwriting from as many relevant people as possible should be examined by the investigator before arriving at a conclusion. By "relevant people" I mean those people who, as far as is known, are involved in the case in one way or another.

Signatures that are suspected of being forged form a special category of handwriting examination. Most commonly, forged signatures are tracings of an original genuine one. Such tracings are often made in pencil and, since the inked-in signatures will not flow consistently over the pencil marks, some of these latter can be detected under microscopic examination, or even by the naked eye. However, more refined techniques are available to establish that a signature has been inked over. Many inks are transparent when viewed under infra-red light, while pencil markings remain opaque. This fact can easily be exploited to examine signatures that may have been forged.

Some forged signatures are not tracings and are done free-hand. These flow more steadily from the writer's hand, unlike tracings, which betray a laboured, hesitant movement over the paper. Nevertheless, even the free-hand forger will hesitate, removing the pen from the paper at intervals, with the result that the end product lacks the fluency of the true signature.

The way in which words are written – as distinct from the style in which they are written – can also yield useful clues. For example, some people write whole words – even long ones – without pausing to lift the pen, while others may pause to lift the pen off the paper in the middle of a long word. Generally speaking, people unused to writing tend to pause in this way; children are particularly prone to doing this when they are learning to join up their letters. Photocopies are of little use in this area, since the pen-strokes cannot be seen on them.

The forging of a whole document is not an easy process and most forgeries are alterations made on existing documents. Often, a word or words are removed by erasing them with a rubber, or by scraping them off with a razor blade. In crudely made alterations of this kind it is usually easy to see the changes with the naked eye, but sometimes such forgeries are done more expertly.

Erasures by a rubber eraser are often not easy to detect by eye. In such cases, lycopodium powder* is used to dust the document while holding it at an angle. The excess is shaken off lightly, but some of the powder will adhere to small particles of rubber that remain on the papers after erasure. Unfortunately, the powder cannot be used to detect the use of plastic erasers.

Some forgers use chemical oxidizing agents or solvents to obliterate words on a document, substituting another word. Such changes are not visible to the naked eye, but microscopic examination will reveal a discoloration in the treated area. Also, if the ink used to write the substitute word is different from the ink used in the rest of the document, this may

Electron micrograph of diatoms.

The assassination of President Kennedy.

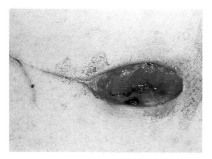

Stab wound caused by a knife.

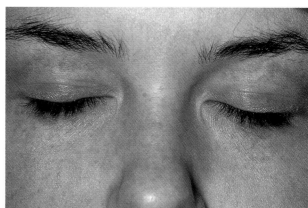

Petechial rash around the eyes.

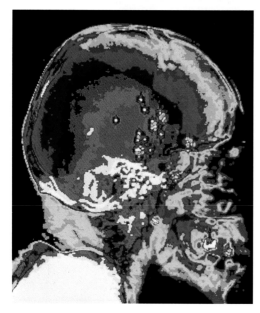

CT-scan showing a gunshot wound to the head (the blue spots are leadshots).

Death Cap mushrooms.

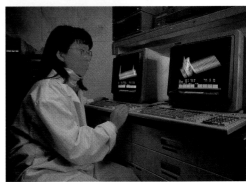

A technician displays
the results of a gas
chromatograph.

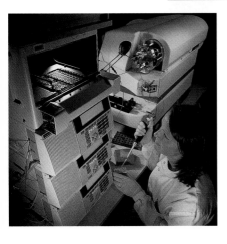

A mass spectrometer.

Rasputin.

The first scene of the Bayeux tapestry. Note that Harold is a full head shorter than his companion.

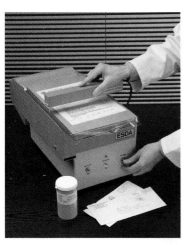

ESDA apparatus.

Accelerant detection apparatus.

Police officers and forensic scientists at a house fire.

Glass refraction apparatus.

Aircraft remains at Lockerbie, Scotland.

Magnified hairs.

Magnified fibres.

Scanning electron micrograph of a piece of glass.

The remains of the bus that was blown up near Tavistock Square in the London
terrorist attacks of July 7, 2005.

be detectable. We have seen that some inks can be rendered transparent under infra-red light; other inks do not react in this way, although they can often be distinguished from one another by the fact that some are infra-red luminescent; when subjected to blue-green light, they will radiate infra-red light, while others will not.

Those inks that absorb infra-red light differ in their ability to do so. Subjecting a document to infra-red light and recording the light reflected from it on to special film may show that different inks had been used. Chromatography can also be used to separate and identify the components of the inks used. A blunted hypodermic needle is used to remove a small sample of ink for laboratory analysis.

Until recently there were no reliable methods of either identifying or of dating inks. With the development of the microspectrophotometer – essentially a microscope attached to computerised spectrophotometer – tiny ink samples can be studied by placing them under the microscope and passing a beam of light through the speck to produce an absorption spectrum. Depending on the light used, it is possible to obtain either a visible or an infra-red absorption spectrum, and this new technique is widely used for examining other kinds of trace evidence such as paint chips and fibres. However, most forensic work done on ink is still concerned with comparing two samples, rather than the identification of a specific make. At least, this is the situation in most countries, including Britain, but the Secret Service Laboratory in Washington, USA, maintains a large collection of inks dating back to the 1920s, which may be compared to inks from forensic cases for positive identification. Also in the USA, some manufacturers now add to their inks fluorescent dyes, which are changed every year, allowing the dating of these inks in the event of their criminal use.

When someone writes something on a piece of paper in a notebook, the paper underneath will, to a greater or lesser extent, be indented. This fact allows investigators to discover what had been written on the top piece of paper if it had been torn out of the book. The indentation may not always be clear enough for decipherment, but a technique known as ESDA (Elesctrostatic Detection Apparatus) can highlight these indentations. Reading such indentations may be important if the writing on the uppermost sheet was, say, a blackmail note or a threatening letter.

ESDA makes use of a physical phenomenon that is not understood scientifically. When a piece of paper is indented, its dielectric properties (i.e. the extent to which it is a non-conductor of electricity) change. Why this happens, no-one knows, but it is useful to the forensic investigator. Essentially, ESDA is an apparatus that releases an electrostatic charge through the paper, which is kept tilted at an angle. A mixture of photocopy toner and small glass beads is sprinkled over the paper. The altered electrical properties in the indented areas will attract the toner-covered beads, making the indentation legible.

Paper is much easier to date and identify than inks, due to the widespread use of watermarks. We have already seen in Chapter 4 how Professor Pokrovsky's forgery of the Yurovsky note was discovered by the simple fact that the paper it was written on was not manufactured until several years after the time the note was supposed to have been written.

There have been attempts to forge watermarks, but these are easily spotted. A watermark is made during the manufacturing process and the image is visible because there are fewer fibres in it than there are in the rest of the paper. Attempts at forging a watermark are concerned with the addition of an image, which can easily be identified as such. The usual methods employed are either the printing of a mark, or the compression of the paper with a bossed design of the mark.

Other attributes of paper include the kind of fibre from which it is made (e.g. wood, cotton, synthetic) and whether or not it had been treated with chalk or fluorocarbons during manufacture, among other things. It is, therefore, often possible to narrow down the field of search quite considerably. Particularly important documents, such as banknotes or passports, are printed on paper that contains fluorescent fibres, special inks and other devices that make the process of forging difficult or impossible.

Charred papers are often met with in forensic investigations, either as a consequence of a fire or the deliberate intention to destroy evidence. Charred bits of paper, as everyone knows, are extremely fragile and may crumble when handled, but there are techniques that can preserve such material. Spraying the fragments with a three per cent solution of polyvinyl acetate in acetone will consolidate them, making them more amenable to flattening without breaking. The pieces of paper can them be floated on a twenty-five per cent solution of chloral hydrate, with ten per cent glycerine in an alcohol solution. The papers can then be photographed under ordinary or infra-red light. The whole process is not an easy one, but in expert hands much of the writing can be revealed. Sandwiching pieces of charred paper between two photographic plates, which are then placed in a dark box for about two weeks, followed by the photographic developing of the plates, is another method of "resuscitating" the writing.

Typewriters, computer printers, photocopiers and fax machines can, in principle, be linked to particular documents, due to the peculiarities of each individual machine, no two of which are exactly alike. Whilst this is, in practice, much easier to do this with an old-fashioned typewriter than it is with electronic machines, nevertheless mechanical wear will produce tiny changes and imperfections. Laser printers, for instance, accumulate small marks on the light sensitive drum and these will appear as minute dots on every document printed by that machine. Photocopiers will replicate these dots, as well as any marks, dirt or dust specks on the glass platen, and the gradual change in these latter marks may even help in dating a document.

FORGERY AND FALSIFICATION

In the Major strychnine poisoning case discussed in Chapter 6, we saw that Mrs Ethel Major accused her husband of being unfaithful, producing letters from alleged lovers and from anonymous accusers. The anonymous letter sent direct to the Chief Constable read: "Arthur Major is allways drunk in charge and not safe on the road. It is my duty to inform you of the above for the safety of the peopel in the village." Note the two spelling mistakes. During one of the interviews with Mrs Major, the investigating officer, Inspector Young, asked her to write the words "always" and "people"; she spelt them "allways" and "peopel".

Simple errors like these can often incriminate a criminal very effectively. Although the deliberate forging or falsification of documentary evidence is usually carried out with a certain amount of subtlety, there are cases in which the record has been altered in the most transparent way. An example of this occurred after the murder of Julie Ward, a young Englishwoman, who was on holiday in Kenya in the eighties.

Julie Ward was murdered in the Masai Mara Game Reserve in southern Kenya in September, 1988. She disappeared early that month and her family in England began to grow concerned about her safety. Her father, John Ward, flew out to Kenya to look for her. He went to the Masai Mara and there had the horrible experience of discovering his daughter's remains himself. All that remained was her jaw and part of her left leg. A fire had been lit nearby.

The story of the horrific murder of Julie Ward has been told many times and I will not go over the whole ground here. Suffice to say that the Kenyan authorities did their utmost to cover up the murder. One way they tried to do this was by tampering with the report on the most important part of the evidence – the skeletal remains.

The leg and jaw were given to Dr Shaker, an Egyptian pathologist in the employ of the Kenyan police. In his post-mortem report he concluded that the remains had been "cleanly cut" with a sharp instrument and that there could be no doubt that foul play was involved. Before the report was released to John Ward, Dr Shaker informed him of its contents, adding, confusingly, that "I shall make sure that it cannot be tampered with".

Puzzled by this odd remark, Mr Ward waited to receive his copy of the report. There were delays. Another visit to Dr Shaker elicited the response that he could not supply a copy of his report, because the police had taken all the copies. Did he not keep a copy for his files? No, replied the doctor, he did not have a copy. Could he type a new report from his notes? No, said the doctor, since he kept no notes! Eventually, through the intervention of the British High Commission, a copy of the post-mortem report was obtained. It had been altered.

Where Dr Shaker had typed that the jaw was "cleanly cut", someone had typed "cracked" over it. The original typing was still discernible and the new typing was misaligned, appearing half a notch below the original. Further on, where the original stated that the left lower leg had been "cleanly cut", someone had typed a series of Xs over the word "cut", erased the word "cleanly", then typed the word "torn". Where Dr Shaker had reported that there was a "sharp" wound on the calf of the leg, the word was altered to read "blunt".

Altered by whom? Before the report reached the British High Commission, Dr Jason Kaviti, Kenya's Chief Government Pathologist, had read it. Accompanied by Dr Shaker, who was in a state of fright, and John Ferguson, of the High Commission, Mr Ward paid a visit to Dr Kaviti at the Kenyatta Hospital. Slapping the report down on his desk, Ward asked the dumbfounded pathologist why he had falsified the report.

No immediate reply came. Eventually, he said that, while he was in the mortuary the previous day, he found the unfinished report and decided to make some "corrections" to Dr Shaker's English, for Dr Shaker's command of the language was not very good. He, Kaviti, simply "knew" that, when Dr Shaker typed "cleanly cut", he really meant to type "torn", so he felt obliged to "correct" the report.

The Kenyan authorities had been claiming that Julie's death had been caused by an attack by wild animals. In view of this, the most startling alteration in the forged report was its conclusion: "Blunt injuries with subsequent burning". In other words, the animals killed their victim, then lit a fire to burn the evidence! If the burning took place after the injuries, then foul play had to be involved. The report had not been altered well enough; it is hard to think of a more unsubtle attempt to falsify the documentary record.

Dr Kaviti was obliged by Mr Ward to initial and date the changes he made, thus admitting his guilt for all to see. As far as I am aware, no action was ever taken against the government pathologist.

A small, but interesting and relevant little event took place during Dr Shaker's investigations of the jaw. Julie Ward's dental charts were obtained from her dentist in England, albeit with some difficulty, since the dentist seemed strangely reluctant to reveal the document. The chart was needed to compare with the teeth in the jaw, in order to confirm that the jaw was, indeed, Julie's. When Dr Shaker examined the chart, he concluded that it matched the jaw and that the jaw did, indeed, belong to Julie Ward. And yet he was slightly puzzled. There was an extra filling on the chart, where there was no corresponding filling in the jaw. Perhaps it was an honest mistake. On the other hand, perhaps the dentist had been charging the National Health Service for fillings he had not performed; this would, at any rate, explain why he was so reluctant to hand over the records.

THE HITLER DIARIES

A more complex case of forgery came to light after the editors of the German magazine, *Der Stern*, announced on April 25, 1983, that they had acquired Hitler's diaries – all sixty-two volumes of them. Although many people seemed to accept them as genuine, forensic investigations soon showed them to be forgeries.

First, the bindings were held in place with polyester threads, a synthetic fabric not manufactured until after the Second World War. In other words, after Hitler's death and after the date of the last entry in the diary. The labels on the diaries were stuck on by means of a glue that contained chemical compounds not in use until after 1945. The paper was of a kind not manufactured until 1955 and tests on the ink suggested it had only recently been applied to the paper.

Other evidence included the fact that it was highly unlikely that Hitler, having suffered injuries during the 1944 attempt to assassinate him, would have been able to write so copiously well into 1945, as suggested by the diaries. His taste for fine book bindings was not reflected in the rather second-rate paper and imitation leather bindings of the diaries.

On the other hand, when the handwriting was subjected to analysis it seemed to match samples of Hitler's writing. However, the origin of these samples was suspect, as they had come from the German Federal Archive's dossier on Gerd Heidemann, the *Der Stern* reporter negotiating the sale of the diary on behalf of his employers. Heidemann had obtained much of his collection of Hitler papers from a dealer, whom it transpired used many aliases but was eventually revealed as Konrad Kujau, the same dealer who had sold him the diaries. The two men were brought to trial, convicted and sentenced to prison terms of four years and eight months.

THE JOURNAL OF JACK THE RIPPER

Another notorious diary discovery of recent years was the alleged journal of Jack the Ripper. It is not a conventional diary, with dated entries, but a rather hysterical screed about the writer's deeds. Alleged to have been written by one James Maybrick – a real historical figure, who was said to have been murdered by his wife – the diary was denounced as a forgery almost as soon as it came to light in 1992, although many people believe it to be genuine.

The debate over the identity of Jack the Ripper is a very complex and involved matter and cannot be entered into here. However, since I am often asked whether or not I believe the diary is genuine, and because the subject seems to be of such widespread interest, I will say a few words about it, restricting the discussion to a brief consideration of the scanty forensic evidence.

Some people maintain that the handwriting does not resemble Maybrick's hand at all, while others say that some of it does, in fact, match his style very closely. As we have seen, a certain amount of subjectivity is involved when forming opinions about handwriting authorship, so this point must be regarded as inconclusive.

Secondly, the ink was found to lack the bronzing that comes with age. Also, samples of the ink dissolved in solvents at the rate one would have expected if it were new ink. As pointed out earlier, dating inks is extremely difficult, but it is, at least, possible to say whether a document was written recently (say, within the last five or ten years or so) or not. Therefore, on the basis of the ink evidence, the balance of probability is that the diary is a forgery, although this can only be looked upon as a temporary opinion, pending better dating methods.

After the diary came to light, a man named Albery Johnson came forward with a watch that was made in 1846. Inside the case the words "J. Maybrick. I am Jack", were inscribed, as well as the initials of the Ripper's five victims. Examination with a scanning electron microscope – an instrument capable of examining the surfaces of objects at very high magnifications – showed that the scratches were old. Also, a blackened particle of brass was found stuck in the scratches; the particle was also deemed old. By "old", what is, of course, meant is that it did not take place "recently"; and something that did not take place "recently" is "old". The circularity of the argument is clear, but it would seem to be the case that the inscription was not made during the past ten or twenty years. Again, the evidence is inconclusive.

So, is the diary genuine, or is it a forgery? The disappointing answer must be "I don't know". Of course, we have considered only the relevant scientific evidence, but there is a vast amount of circumstantial and historical evidence to be considered. The reader who wishes to pursue this matter should consult the books listed under "Further Reading".

It is not only the written word that can be used as evidence. Images of all kinds can help us arrive at the truth. After all, written words are but images of a certain kind, and we must not forget that the earliest "writings" were series of pictures.

THE TURIN SHROUD

"Someone got a bit of linen, faked it up and flogged it". With these words, one of the most famous images to exercise the attentions of forensic scientists – the image on the Shroud of Turin – was dismissed by Professor Edward Hall of Oxford University. His comment came after his, and two other, laboratories arrived at the conclusion that the Shroud was a forgery made sometime between 1260 and 1390. The conclusion was based on radio-carbon dating methods.

Prior to this finding, analysis of pigment particles from the Shroud showed them to be red ochre and vermilion with collagen tempera medium, a paint composition common in the fourteenth century. In fact, the Shroud first came to light during that century, when it was in the possession of a French soldier, Geoffrey de Charny. After that, the Shroud changed ownership several times; by 1983, ownership had passed to the Pope, although the Shroud itself is kept in Turin.

So, carbon dating and pigment analysis indicated a fourteenth century origin. In addition, it is interesting to record that the French church denied the authenticity of the Shroud when it first appeared. In fact, one bishop went so far as to write to the Pope, saying that he knew the artist who painted it.

Is this, then, the end of the matter? Has science disproved the authenticity of the Shroud? The startling answer is that the matter is as far away from resolution as ever. Let us first take a look at the Shroud itself. It is a length of linen cloth, measuring 437cm long by 111cm wide. It is woven in a tight herringbone pattern and it bears the front and back images of a person. The image is not at all clear; the well-known published photographs of it are, in fact, negatives, which show the image more clearly than do the developed prints. Finally, the images are said to be those of Jesus Christ and the cloth was the shroud in which he was wrapped after being brought down from the cross.

What supports this last assertion? First, the herringbone weave is historically unusual and does not resemble any known mediaeval fabric. On the other hand, there examples from Egypt, dating from Ptolemaic (i.e. Greek) times. The first European examples of this kind of weave date from late in the sixteenth century. The stitching, too is unusual, but it does resemble the stitching in some textiles from Masada, the Jewish fortress destroyed by the Romans in AD73. The shape of the Shroud – long and narrow – is also very unusual, being extremely rare at any time in history and it was certainly not typical of fourteenth century France.

It has not been established how the image was made. There was only a minute amount of pigment on the cloth. No outline is present around the image, which, in all probability, would have been the case if it had been done in the Middle Ages. In order to reproduce a similar image, modern science has had to resort to the use of techniques that were unavailable to mediaeval artists. The images most closely resembling the ones on the Shroud were produced by injecting a person with a short-acting radioactive isotope, then wrapping him in a cloth.

The Shroud was kept folded over for several centuries, yet the image did not soak through, as would have happened had the image been painted. Traces of blood were found on it, but, of course, no-one can tell whose blood it was. The image itself is strange; it seems to represent, not a man lying on his back in a normal position, but a man in a contorted position, with legs bent – a position that is very difficult to hold for any length of time, as would have been required from an artist's model. It has been suggested that the position is similar to that of a body, fixed in rigor mortis, that has been brought down from a cross.

Of particular interest is the pollen that was extracted from the cloth. The spores belonged to two plant species: *Zygophyllum dumosum*, which has a distribution restricted to the Jordan Valley and the Sinai Peninsula; and *Gundelia tournetfortii*, which is also restricted to the Middle East, although it has a wider distribution in the region. Small pieces of wood were found in the region of the image of the back of the head, and the suggestion has been made that these came from the cross itself.

Although it is true that there is no verified reference to the Shroud prior to the fourteenth century, earlier references to a shroud of the same kind are known. Sixth-century references to a cloth in Edessa (modern Urfa in Turkey) describe it as having the face or body of Jesus miraculously imprinted on it. It is said that this was taken to Constantinople in 944, but it disappeared from there after the sack of the city by the Crusaders in 1204. Interestingly, a manuscript in Budapest, dating from the 1190s, shows the body of Jesus being wrapped in a shroud that is very similar to the Shroud of Turin. The herringbone weave, as well as several distinctive tears to the cloth, suggest that the picture is of the Shroud of Turin itself.

If the Shroud had been forged, to what end would it have served? Geoffrey de Charny was a poor man and an undistinguished soldier; and could not have afforded the price a forger would have asked. How he came to have it is a mystery, but it is possible that he was related to a Templar Knight of the same surname, who was put to death by King Philip IV, when he dissolved the Order of the Temple. If so, then – speculating wildly – Geoffrey may have inherited it from his more distinguished ancestor, who may have brought it to France from the Holy Land, if that is where it did, indeed, originate.

The evidence as a whole does seem to point to a pre-mediaeval, Middle Eastern origin, yet the radiocarbon dating and paint analysis results point the other way. In my view, the pigment analysis means very little, since the paint could have been added later – there is nothing to indicate that the pigments were present on the cloth at the time it was made. As for the radiocarbon results, the author Ian Wilson has suggested that bacterial and other contaminating matter may have confused the issue, resulting in a more recent dating. This kind of error has been known to happen in other cases.

THE BAYEUX TAPESTRY

"*Hic Harold Rex interfectus est.*" The inscription on the Bayeux Tapestry cannot be more explicit. "Here King Harold is killed." Beneath the words "*Harold Rex*" stands a soldier in helmet and mail armour, carrying a long shield on his left arm. His right eye appears to be pierced by an arrow, whose shaft he is gripping with his right hand. There can be no doubt about it; as every schoolchild knows, King Harold II was killed with an arrow that pierced his eye at the Battle of Hastings on October 14, 1066.

But did this really happen? Some historians, notably Sir Frank Stenton, had their doubts. The Bayeux Tapestry is the Norman account of the events leading up to and including the Battle of Hastings. Its basic argument is that the invasion of England was justified, because Harold, having sworn an oath to William, Duke of Normandy, acknowledging him as the future King of England, broke his solemn promise and proclaimed himself king after the death of King Edward the Confessor, who himself had wanted William to succeed him.

The Bayeux Tapestry is essentially a work of propaganda.** It seeks to justify William's invasion, invoking Harold's oath and Edward's blessing. But it is debatable whether King Edward had any right to offer William the crown, since the succession was settled by a council decision in those days. Also, Harold made his oath when he was William's prisoner, after he was ship-wrecked off the French coast, probably in 1064. He may have felt compelled to take the oath on pain of permanent imprisonment. Furthermore, William asked Harold, while taking his oath, to place his hand on a box, which contained sacred relics. It is said that William, deceitfully, did not tell Harold that the box contained these relics and, therefore, he (Harold) may not have realized that he was swearing a solemn oath, but merely making a promise under duress. Nevertheless, William used the incident to discredit Harold, denouncing him as a man who broke his oath, sworn in the name of holy relics.

When King Edward died, early in January, 1066, Harold proclaimed himself king. Outraged, or affecting to be outraged, William invaded England and defeated the Anglo-Saxon and Danish forces on October 14. It is certain that Harold was killed during the battle.

But how was he killed? I am not quite sure what made me change my mind about the conventional version of history. Somehow, the accepted story of Harold's death did not seem convincing when first I saw the Tapestry at Bayeux. Something was definitely wrong, but it was not immediately apparent what it was. This is particularly odd, since the visitor looking at the Tapestry is informed by a voice on a taped commentary that the figure with the arrow in its eye is Harold II; no doubt whatsoever is expressed. However, I was aware that there were objections in scholarly circles to the schoolbook account of Harold's death and I decided to find out what these were. I consulted a number of historical works on the Tapestry and the Battle of Hastings, as well as contemporary documents, and I discovered that there were three reasons for the doubts that some scholars had expressed.

First, with one exception, there is an almost complete absence of any early documentary evidence suggesting that King Harold was shot dead by an archer. The *Anglo-Saxon Chronicle* says only that Harold was killed during the Battle of Hastings and Snorri Sturluson, the Icelandic saga-writer, in his monumental *Heimskringla* (*The Orb of the World*), says the same. Neither the *Anglo-Saxon Chronicle* nor the *Heimskringla* mentions an arrow. In the case of the *Heimskringla* it is highly probable that Harold's death by an arrow would have been mentioned had it been true, for Snorri Sturluson was fond of drawing parallels between the Battle of Stamford Bridge, in which Harold defeated King Harald Hardrada of Norway, and the Battle of Hastings, in which Harold himself was defeated. Sturluson records that Harald Hardrada was killed when an arrow struck him in the throat at Stamford Bridge; had he known that Harold died a similar death he would almost certainly have recorded it in his saga. In matters like these, what is not said is often as illuminating as what is said. Finally, and most revealingly, Guy, Bishop of Amiens, wrote that Harold was brought down by four Norman knights: Guy, Count of Ponthieu; Eustace of Boulogne; Walter Giffard and Hugo de Montford.

Behind the alleged "Harold" figure in the Tapestry there is an illustration of a Norman knight striking down with his sword a soldier on foot. The figure thus struck down is shown dropping a long-handled battle-axe and above him appears the second part of the inscription, "interfectus est" – "is killed". It has been suggested that this second figure is meant to depict the real Harold.

The only early written reference to King Harold being killed by an arrow was made by Baudri, the Abbot of Bourgeuil, at the turn of the twelfth century, in other words, about thirty-five years after the Battle of Hastings. The Abbot wrote a poem in which he referred to an embroidery, which, in all probability, was the Tapestry itself, and therefore his comment is very likely to have been the result of a misreading of the Tapestry, not a genuine record of historical observation.

The second reason for scholarly doubt is that there appears to be an invariable rule in mediaeval illustration that every violent death is indicated by a falling, bent or prostrate figure; the alleged figure of Harold in the Tapestry is standing upright.

The third and final reason given by historical scholars for doubting the authenticity of the "Harold" figure concerns another illustrational convention, which asserts that a single named individual never takes a secondary place in any action; the arrow-in-the-eye figure is said to appear as such a secondary figure in that particular sequence of the Tapestry.

I must confess that I do not find the second and third reasons at all convincing. The Tapestry is, in many ways, a unique historical record and I see no reason why one should impose artistic rules upon it. It was probably made in an Anglo-Saxon workshop on the orders of Odo, Bishop of Bayeux, who wanted to hang it in the cathedral at Bayeux to glorify his own part in the Norman Conquest. It was more of a political than an artistic piece of work. Under such circumstances, the niceties of artistic conventions do not seem to me to carry much weight. Also, I cannot see what is meant by a secondary figure; both the arrow-in-the-eye figure and the figure being knocked down by the Norman knight seem to me to be roughly equivalent in the action portrayed.

On the other hand, the contemporary documentary evidence is very compelling. The Bishop of Amiens is very precise: four named knights brought Harold down. No eyewitness or other contemporary person suggested that Harold died any other way. The presence under part of the inscription of a figure being struck down by a Norman knight suggests that that figure was meant to be Harold.

Another source of confusion has crept in. The Tapestry was damaged during the French Revolution and was nearly destroyed, but it was rescued in 1792 at the last minute by Lambert Léonard-Leforestier, a lawyer and member of the Bayeux City Council, from soldiers who were using it as a wrapping cloth to cover the goods in a horse-drawn cart. Many parts of the Tapestry were repaired and it is said that the figure with the arrow in its eye was one of the sections that were thus repaired. Some say that there was no arrow in the eye of the original figure. If this is true, then the account given by the Abbott of Bourgeuil cannot be dismissed, because he could not then have got the idea of the arrow in the eye from the Tapestry, as has been suggested. However, for reasons that will become clear later, I do not believe that the arrow in the eye is a recent addition.

When I asked myself why I did not believe that the arrow-in-the-eye figure was Harold, I could only give myself the bizarre answer that he did not look like Harold! By this I mean that the figure simply did not look like all the other undoubted representations of Harold in the Tapestry. Of course, one cannot expect a photographic likeness in an embroidery of this sort, but the figure appeared to me to differ so much from the others that I decided to write down a description of it to compare with the others.

The most obvious attribute of the alleged Harold-with-an-arrow-in-his-eye figure that strikes the onlooker is that it is large by the general standards of the Tapestry, although it is by no means the largest figure. Conversely, Harold is always depicted as a small man and is always shown to be smaller than William or Edward the Confessor, despite the fact that, as a Dane, he was, in all probability, taller than either of them. William is known to have been a short man and, while there is no record of Edward's height, he was a very old man at the time the events in the Tapestry were meant to have taken place and it is very unlikely that he would have been taller than the Viking Harold. Indeed, Harold was known to have been a particularly well-built and powerful man; even in the Tapestry he is shown dragging a soldier out of a sandbank and carrying another over his shoulder, a feat only a very strong man could accomplish. Nevertheless, the figure of Harold in that scene is one of a small, very slightly built man.

The best way to describe the size of a figure is, of course, to measure it. I decided to do this to the alleged Harold figure, as well as to the many certain representations of Harold and other characters shown in the Tapestry. The height of the arrow-in-the-eye figure, which, for the sake of ease of reference, I will call X, is sixty-five per cent of the full height of the Tapestry. The undoubted Harold in other scenes never reaches that height.

In the first scene of the Tapestry King Edward, in spite of being seated, appears to be a veritable giant when compared to Harold, who looks almost like a child. Even Harold's companion (a nameless figure) towers a full head above him. In that scene Harold's height is fifty per cent of the Tapestry's height. In the scene in which Harold and his companion enter Bosham Church the height of Harold (adjusting for his bent knee) is about forty-two per cent of the height of the Tapestry. In the scene in which he is arrested by the Count of Ponthieu, Harold is fifty-six per cent the height of the Tapestry. When Harold is interviewed by the Count, the latter, although seated, appears much larger than Harold, who is forty-four per cent of the Tapestry's height. In the scene showing Harold and William at the Palace in Rouen, William is as tall as Harold, although he, like the Count of

Ponthieu and Edward the Confessor in earlier scenes, is seated; it is difficult to measure Harold in this scene, because he is depicted in a semi-crouched position. In the dramatic scene in which Harold takes the Oath of Allegiance to William, William, again seated, is both taller and bigger than Harold, who is about fifty per cent the height of the Tapestry. When Harold returns to England and is received by King Edward, who is again seated (he never appears standing anywhere in the Tapestry), the king is depicted as a giant in comparison with Harold. If one imagines the figure of Edward standing, he would be almost as tall as the entire height of the Tapestry; this also holds true for the seated figure of William in the Oath Scene. In the scene in which Harold is received by the king, Harold is fifty-one per cent of the Tapestry's height, although the position of his head, bent in shame, makes it difficult to make a reasonable estimate of his height. It seems clear, however, that he is deliberately being depicted as a much smaller man than the king. When Harold is offered the crown he is fifty-nine per cent the height of the Tapestry – the largest figure of Harold (not including X) in the entire embroidery. Nevertheless, even here he is shown as being shorter than the figure of the man who is offering him the crown. Even when crowned and enthroned, Harold is shown as a considerably less impressive figure than either Edward or William. In the second scene showing Harold on his throne, he is depicted as a cowering and bowed figure, terrified by the arrival of an evil omen in the shape of a flaming star. In fact, this star was Halley's Comet, which appeared in the spring of 1066 and which is shown in the top margin of the Tapestry. Several people are looking up and pointing at it, while a messenger rushes to tell Harold of this ominous sign.

I have not measured all the Harold figures in the Tapestry, since some are represented in a manner that makes it difficult to measure the figure meaningfully, such as representations of Harold mounted or when he is rescuing two soldiers from the River Couesnon.

King Edward the Confessor and William, Duke of Normandy, are represented as good and honourable men in the Tapestry; Harold is meant to be the villain. It would seem reasonable, therefore, that he would be depicted as a symbolically smaller man than the other two. William hardly ever appears standing in the Tapestry, with the exception of the scene in which he is bestowing arms upon Harold after the defeat of Conan, Duke of Brittany. (A standing figure of a knight being brought a horse appears after the scene of houses being burnt at Hastings, but it is uncertain whether this is a figure of William; it measures sixty-six per cent of the

Tapestry's height.) In the arms-bestowing scene, the only one in which William and Harold are shown standing together, William is not shown as being very tall, but he is shown as being taller than Harold; his figure is fifty-eight per cent of the Tapestry's height, while Harold's is only fifty-three per cent. Furthermore, William is shown to be a generally more robust man. William's head is always larger than Harold's, whereas X's head is very large and the whole figure is of a very robust man. Harold is generally represented as being of slight build – compare, for example, X's build with the figure of Harold rescuing the two soldiers from the river.

Apart from the discrepancy in size and build between X and the undisputed representations of Harold in the Tapestry, two other pieces of evidence that suggest that X is not Harold may be invoked. The first piece of evidence lies in Harold's facial appearance. While it would be absurd to claim that the embroidered representations of the faces of the various characters are accurate likenesses, nevertheless, the various characters in the Tapestry do have a consistency of appearance. For example, the old, bearded and crowned King Edward is always easily recognizable and William, with his clean-shaven face and distinctive hair-style, is also recognizable when he appears. And so is Harold, who is generally depicted as having a small face. If one compares X with any of the undoubted Harold figures it is easy to see that he is not the same character. On the other hand, the falling figure under the words "interfectus est" is very similar in facial appearance when compared, for example, with the figure of Harold in the Oath Scene.

There is another piece of evidence that goes against the idea of X being Harold. As we have seen, the Tapestry was intended to be a piece of propaganda supporting William's claim to the English throne and justifying his invasion of the country. Harold had broken his oath, sworn on sacred relics. He had to be punished, but defeat and death in battle, it seems, were not punishment enough, for in later years the story was embellished. Some time during the twelfth century, the story arose that Harold was killed by an arrow in the eye as a sign of divine retribution.

Interestingly, this story weakens the case for X being Harold a good deal. The reason is this: X is the third figure in a series of three soldiers standing in a row and he has an arrow in his eye. Slightly earlier in the Tapestry and lying in the break in the word "*cecide————runt*" there appears a similar row of soldiers, the third of which also appears to have an arrow in his eye. It would seem to be highly unlikely that both these figures were meant to represent Harold, since there is no mention of Harold at all in that part of the Tapestry in which the first arrow-in-the-eye figure (not X) appears. Behind the first arrow-in-the-eye figure there appears another figure falling over with an arrow in his eye and in the margin below these two figures there appears a third, dead, figure also with an arrow in his eye. Furthermore, at the very end of the Tapestry there appears a fleeing figure with an arrow in his eye, immediately behind a mounted figure who also seems to have what looks like an arrow in his eye. Therefore, the suggestion that the arrow in the eye of X was meant to be symbolic of divine retribution visited upon Harold for breaking his oath also appears to be unfounded, being diluted, so to speak, by the abundance of other arrow-in-the-eye figures. In other words, the arrow in X's eye has no special significance; arrows in the eye are two a penny.

The frequency with which the defeated Saxons are shown with arrows in their eyes also suggests that the arrow in X's eye is not a recent addition, sewn on when the Tapestry was repaired in the nineteenth century, but that it was embroidered by the Tapestry makers themselves. Also, the position of X's right arm, raised toward his head, as well as the depiction of his right hand, suggest that he was meant to be shown gripping something; if the arrow had not been there in the original, it is difficult to explain the position of the arm and hand. Finally, the arrow cannot really be seen to pierce the eye at all; it appears to enter the head in the region where the right eye should be, but the figure of X stands in left profile and it is in our imagination that we see the arrow piercing the eye. Nowhere does the Tapestry say that an arrow pierced X's eye. It follows from all this that the Abbot of Bourgeuil's belief that Harold was killed by an archer was based on a misreading of the Tapestry.

All this led me to conclude that King Harold II was not killed by an archer and that the contemporary account by the Bishop of Amiens is the correct one. The falling soldier under the words "*interfectus est*" is almost certainly meant to be Harold.

But is it strictly true to say that there is no evidence supporting the schoolbook story? In the Tapestry the words Harold Rex appear directly above the X figure. Is this not good evidence that the old story may be true after all? No, it is not, for there are several scenes in the Tapestry in which Harold's name appears above representations that are indisputably not of Harold. In some instances, his name appears some distance away from his pictorial representation. In fact, once all the evidence is examined, there is nothing at all to support the claim that King Harold was shot in the eye with an arrow.

THE SPOKEN WORD

It is not only the written word or image that can point to the truth; the spoken word can also yield evidence. Take this fairly simple example. On June 13, 1994, the California police caught up with O.J. Simpson at his hotel in Chicago and an officer informed him that his ex-wife had been murdered. Immediately, Simpson became hysterical; he wailed and lamented, while the police officer watched him with increasing bewilderment, for something was wrong. The officer suddenly realized that he had not told Simpson which one of his two ex-wives had been killed, for Simpson had been married twice. Simpson did not ask which one of his ex-wives was the victim; or how or where or by whom she was killed. In fact, he asked no questions at all.

This example may not deserve the epithet "forensic", but it is such a telling instance of how damning the spoken word can be that I could not resist mentioning it. On a more scientific level, attempts have been made to describe the individual human voice objectively, with the aim of identifying a person. These days threatening or obscene telephone calls seem to be increasing and, if these can be recorded and later compared with the voice of a suspect, an arrest could be made if the voices are a "match". The greatest advances in the technology required are being made in the USA. In Britain and Europe, voice identification is still in its relative infancy.

We can all identify the voices of our friends and relations, although we may sometimes make mistakes. Nevertheless, there is a prima facie case for believing that voices ought to be as distinctive as many of our other attributes: our gaits; our faces; our fingerprints; our DNA. It was on the basis of this belief that the idea of a voice spectrograph was developed in the late 1960s.

Initially there were two types of spectrograph, each producing a different kind of voiceprint. The Bar Voiceprint recorded a 2.5 second segment of speech on magnetic tape. This was then electronically transferred to a stylus, which made a trace on chart paper on a revolving drum. The resulting chart looked like a bar code fluctuating up and down. The denser the print, i.e. the closer together the bars, the louder the voice, thus:

Quiet voice: | | | | | | | | Loud voice: ||||||||||||||

The general shape of the print was also said to be unique to an individual. Thus, two different voiceprints could be represented as follows (these representations are diagrammatic):

|||||||||||||||||||||||||||||||||||||||||||||||||||||||||||||||||||||||||||||| and

||||||||||||||||||||||||||||||||||||||||||||||||||||||||

Another kind of spectrograph produced a contour voiceprint. These looked like the contour lines on a geological map and were said to be unique to the individual. With the advent of digital signal processing, electronic sound spectrographs and sonographs (essentially sound wave analysers) have been developed that can display recorded speech in three formats, time, frequency and amplitude, on a computer screen, although so far these have mainly been used for analysing animal sounds.

I hope that the cautious manner in which this section is written is not lost on the reader. New techniques of acoustic analysis are being developed, especially ones that can be carried out by computer programmes, for example measuring a range of voice parameters such as the rate of vocal chord vibrations, which produce the pitch. However, in this area of voice analysis, forensic scientists are still groping towards the light. Having said that, some practitioners in the USA have claimed considerable success in voice identification. For example, when Clifford Irving claimed to be publishing the autobiography of the recluse millionaire, Howard Hughes, a group of press, radio and television reporters met at Los Angeles, where a television company asked for a comparison to be made between Hughes' voice, as heard on the telephone, and an undoubted earlier recording of his voice. This showed that the refutation made by Hughes was, indeed, made by him, not an impostor, and Irving was later found guilty of forging the so-called "autobiography".

Another approach to voice identification is the technique of narrowing down suspects on the basis of regional accents. In Britain, Dr John French is a veritable Professor Higgins in this regard and has had a number of successes in identifying people. He employs what he calls "isoglosses", which are wavy lines on a map, showing where a particular accent merges with another. This is too specialized an area to discuss further here, but it seems to me to be a subject with a promising future.

THE POLYGRAPH

One of the most famous pieces of forensic equipment, the polygraph or lie-detector, is also one of the least reliable. It must be stated at the outset that the polygraph cannot detect whether someone is lying or not. It can only tell us whether the person being tested is agitated or not. It gives us information about the physiological and psychological condition of the person, but it cannot tell us why the person is in that state. For example, a shy person may react adversely to the machine, when a more outgoing person may not. Similarly, a sick person may perform badly, although he may have done better had he been well. So, the machines "conclusions" will differ, not only from person to person (irrespective of whether they are telling the truth or not), but between different occasions for the same person.

Although the machine continues to be in use in the USA, the American government has issued a list of the kinds of people who are unsuited to testing in this way. These include people who are upset or tired, people who are hungry or thirsty, people who are drunk or drugged, habitual liars, demented people, prisoners, children, people taking medication, and so forth. It occurs to me that the list includes almost everybody!

A more recent innovation in the search for the truth in the spoken word is the voice analyser. Unlike the polygraph, the person tested does not have to be attached physically to the machine. It is said that sub-audible tremors during speech can be detected by the instrument; these tremors being, apparently, characteristic of people who are lying. Several similar devices are being marketed in the USA, but I have to say that I view these developments with deep suspicion.

Finally, I must mention a story told me by Dr William Aulsebrook of South Africa. Apparently, the Zulu people have a fail-safe method of identifying a liar. The suspects are brought before a judge and each is fed a spoonful of flour, then asked to speak. Only the guilty man will have a dry mouth and the flour will fly about out of his mouth as he speaks, revealing his guilt. I cannot say that the system is in any way inferior to the polygraph; the reasoning is the same, but the conclusions are equally unreliable.

---

* Lycopodium powder is composed of the yellow spores of the club moss, *Lycopodium clavatum*. It is used in the manufacture of fireworks and certain cosmetics.

** The Bayeux Tapestry is not, in fact, a genuine tapestry, but an embroidery.

CHAPTER NINE

LITTLE DETAILS

Trifles make perfection – and perfection is no trifle.
Michelangelo

I was sitting on a chair in an otherwise empty room. I was at a "scene", making notes about the case. Presently, the police inspector walked in, carrying a small plastic bag containing some soil and debris. He handed it to me and I held it up to the light. I handed the bag back to the inspector and told him that he was right – the man he suspected was, in all probability, lying.

The inspector smiled contentedly; his detective's instincts had not let him down. The few insects among the debris suggested that the body of the murder victim, who had been found under the floorboards of another room in the house, had been moved there from another place, in contradiction to the suspect's claim. It was all very simple.

Of course, my initial impression in this case had to be followed up with more detailed work in the laboratory, but it was the fact that the insects were so out of place where they were found that presented the first, dramatic clue.

It is easy to trivialize trifles. But, so often in forensic work, as in life in general, it is the minute little details – a scratch here, a smudge there – that are the most revealing. The smallest thing, if it is out of the ordinary, can tell us a great deal. But the Devil, as they say, is in the detail.

This chapter deals with those little pieces of evidence that keep cropping up in forensic cases – the sort of evidence with which Sherlock Holmes would have felt comfortable. One of the most frequently encountered pieces of such physical evidence is shattered glass, which is often found at the scene of a burglary, where a window was smashed, or at a murder, in which a bottle-fight resulted in death, or a shooting, where bullets were fired through a window. Many other situations can be imagined.

TWO USEFUL PROPERTIES OF GLASS

Glass has two very useful properties, from the forensic point of view. It refracts light and it has a certain density. Both these qualities can be measured and used as a means of identifying the fragments, i.e. to arrive at a conclusion about the origin of the glass – did it come from the smashed window at the site of the burglary? Did it come from the bottle found broken at the scene of the murder?. Often a suspected criminal will have fragments of glass adhering to his clothes or shoes; these can be compared with glass from the scene.

Refraction is the apparent bending of light when it moves from one medium into another. The most commonly experienced example is when a stick is immersed in water, making it appear as though it is broken at the point of entry. This happens because the speed with which light travels changes from medium to medium. The ratio of the speed of light in air to its speed in water is termed the refractive index of water. The speed of light through air is about 300,000,000 metres per second, but when it enters water it is slowed down to about 225,000,000 metres per second. Therefore, if we take the refractive index* of air to be 1, then that of water is 1.33, since:

$$300,000,000 \text{ divided by } 225,000,000 = 1.33$$

In glass, light is even more highly refracted since it moves more slowly through it – about 200,000,000 metres per second – than through water. Consequently, the refractive index is higher, being around 1.5. The exact refractive index of any particular kind of glass will be slightly above or below this figure.

The refractive index of a glass fragment can be measured by immersing it in a drop of silicon oil and examining it under a microscope with a hot stage, i.e. a stage that can be heated slowly at a known rate. As the oil warms up, it's refractive index will change, until it reaches a point when it equals the refractive index of the piece of glass. At this point it will disappear from view. Since the refractive index of the oil at different temperatures is known, the investigator can conclude that the refractive index of the glass is equal to that of the oil at the temperature at which the glass became invisible. By linking a video camera and computer to the hot stage microscope, it is possible to automate this process, since the camera can record contrast changes as the oil is heated. Once the match point is reached, i.e. when the glass sample has the same refractive index as the oil at that particular temperature, then using stored calibration data, the match point temperature is converted to a refractive index.

The refractive index is not a unique identifying feature of glass, since most glasses have very similar indices. If the glass from, say, the clothes of the accused is compared with glass from the scene and found to be different, the accused may be eliminated from the inquiry. But if they are the same, it is no evidence of guilt; it simply means that we have to go further into the matter.

The density of the glass fragments from scene and suspect can then be determined. This is done by exploiting the fact that solid particles will float, sink or remain suspended in a liquid, according to whether they are less dense, more dense or equal in density to that liquid. The liquid most commonly used forensically is a mixture of bromoform and bromobenzene. The glass fragment (say, from the scene) is placed in this mixture and further amounts of bromoform or bromobenzene are added until the fragment remains suspended in the liquid. The glass fragments from the suspect's clothing can then be immersed in the liquid to see whether they, too, will remain suspended. If they do, then both scene glass and suspect glass have the same density; if they float or sink, the two groups of fragments are clearly different in density.

Let us say that the two sets of glass we are comparing have the same refractive index and the same density. This will strengthen our belief that the two sets had the same origin, but it would still not be conclusive. Certain other examinations can be carried out, based upon the chemical composition of glass. Essentially, glass is made by heating a mixture of sand (silica or silicon dioxide, SiO_2), limestone (calcium carbonate, $CaCO_3$) and soda (sodium carbonate, Na_2CO_3). However, glass also contains various impurities, which are either deliberately or accidentally added during manufacture. For example, sheet glass often contains four per cent magnesia (magnesium oxide, MgO), while bottle glass usually contains about two per cent aluminium oxide (Al_2O_3). Iron salts often occur in glass, but these are eliminated from the glass used to make milk bottles since they add a greenish tinge to the glass, which makes the milk look unappetizing! (The iron salts are what make sheet glass look greenish when look at side on.)

The presence of such impurities can be detected by the use of scanning electron microscopy, which uses a beam of electrons, rather than light, to magnify an object (see previous chapter). Objects under the focus of an electron beam will emit radiations that are unique to the chemical elements contained in it. These chemical signatures can be isolated and monitored through the use of energy dispersive X-ray analysis, a technique that allows the determination of both the presence of certain elements and their respective amounts in the specimen. A more destructive method, emission spectrometry, can be used to detect the presence of trace elements in glass. Here, the material is essentially burnt using a high-energy laser pulse and the spectra emitted used to identify the elements. However, use of this method means that the item is destroyed and cannot be used for further forensic investigations.

While the above techniques depend on the use of sophisticated apparatus, some simple, common sense procedures are used in the examination of glass. If it is coloured glass, then the two samples in question can be compared by eye. It is also possible to determine whether the glass came from a flat sheet, or from a curved object like a bottle, simply by examining it. Slightly more complex is the matching of pieces of glass together, which can be done if fairly large pieces of glass are found. Glass fragments that appear as though they might fit together can be so fitted by hand, and perfect fits will hold together very tightly, strongly resisting separation.

The identification of a piece of glass as having a common origin with another piece of glass thus depends upon the number of attributes that both sets share and on the rarity of some of the attributes. For example, a glass with a particular refractive index may be so rare that the probability of finding two sets of fragments having the same index, but which do not have the same origin, must be regarded as being remote.

It is not only the fragments themselves that can yield useful information. The way in which a piece of glass broke can also be used as evidence. Consider, for example, what happens when a bullet – or a projectile travelling at very high speed – is shot through a window. The glass molecules at the point of impact will dislodge the molecules in front of it; this continues until the bullet emerges through the other side of the pane. The resulting hole will be crater-shaped, the larger end being at the point of exit of the bullet, the smaller at the point of entry, thus:

Of course, the presence of glass particles on the floor on the side where the bullet emerged will often betray the direction of the bullet, but such fragments can be very small and, in any case, they may have been removed in an attempt to conceal evidence.

If a projectile travelling at low speed hits a window pane, the crater-shaped hole will not form, but a star-like or radial pattern of fractures emanating from the hole will be evident, thus:

Circular, arc-like fractures can often be seen around a hole, connecting some of the radial fractures, thus:

These fractures will form on the outside of the window, i.e. the side struck by the projectile, in the following way:

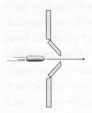

It is difficult to distinguish between a hole made by a high-velocity bullet and one made by, say, a pebble travelling at high speed. A window struck by a slow-moving stone will shatter in its entirety, as will one that is subjected to a close-range shot. However, the presence of gunpowder residues on the glass fragments will often indicate that a firearm was used.

Of course, there is a gradation between the crater-shaped hole and the hole surrounded by radial and circular fractures. The latter can be caused by a bullet if it is travelling at lower speed.

Sometimes two or more bullets may pierce a window and, if radial fractures are present, it is possible to determine which shot was fired first. This is because the radial fractures caused by the second shot will end at the existing fractures caused by the first shot, thus:

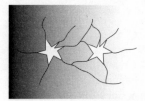

It is clear that the hole on the left resulted from the first shot, the one on the right having been made later.

THE NATURE OF HAIR

Hairs, both human and animal, are often found at crime scenes. In principle, they ought to be very good evidence, routinely used, but, in practice, this is not the case. We know that human hair differs from animal hair and that the hair of different animal species differ from one another. Also, human hairs differ from person to person and from one part of the body to another. Unfortunately, few systematic studies have been carried out, the results of which would have enabled forensic scientists to use hairs routinely as a means of identification, or as evidence that a particular person had been at a scene. Another problem is that hair shows great morphological variation, even within the same individual. Of course, this does not mean that we cannot make more general statements on many occasions. For example, the head hairs of black people and those from white people are quite easily distinguished from one another; and animal hairs are easily are very different from human hair. The hairs of black and Far Eastern people, although they differ greatly between the two groups, are remarkably uniform within each group. Conversely, the range of variation among people of European origin is very great. As a result, in most cases in which hair is used as evidence, the investigation involves a comparison between hairs found at the scene and those from a suspect.

The structure of a hair is quite simple. It consists of an inner core, the medulla; a surrounding layer, the cortex; and a covering "skin", the cuticle. The British forensic scientists Angela Gallop and Russell Stockdale have compared this structure with that of a pencil. Thus, the lead represents the medulla; the wood the cortex; and the paint the cuticle. Sometimes, however, the medulla is absent. When it is present, it may occur in a continuous, uninterrupted line along the length of the hair, or it may be interrupted or fragmented in various patterns, thus:

| ———————— | ——— ——— ——— | ——— —— ——— |
| Continuous | Interrupted | Fragmented |

In most human races the medulla is rarely continuous, although it is the norm in people from the Far East (China, Japan and South-east Asia). The shape or pattern of the medulla can also be distinctive and can readily be seen under the light microscope. Another attribute of a hair is its medullary index, which is the ratio of the width of the medulla and the width of the whole hair shaft. Typically, the figure is less than 0.3 in humans, whereas in most animals it is at least 0.5.

The hair cuticle is not smooth like a continuous skin, but arranged as overlapping scales, like those of a fish. Again, the pattern of these scales can be of diagnostic value. These and other characteristics of hair should, one day, allow us to produce a complete atlas of hair forms, both human and animal, but, until this is achieved, most investigations will have to depend on comparisons made on a case-by-case basis. This is, perhaps, the moment to mention that work-horse of the forensic scientist's armoury – the comparison microscope. This is simply a microscope that enables one to view two objects simultaneously, with the purpose of comparing them. This is used to compare all sorts of items, from bullets to pieces of vegetation, but it is perhaps fair to say that nowhere does it come into its own as when it is used to compare hairs and other fibres.

Hairs found at a scene can yield other kinds of information. For example, it is possible to say whether a hair was forcibly removed from its owner or not. This is because a hair torn out of the scalp will have pieces of tissue adhering to the root, whereas hair falling out naturally will have a root, but no adhering tissues. Hair roots from forcibly removed hairs can be used as a source of nuclear DNA for fingerprinting, because in these cases the hair was still in a growing phase. Naturally shed hair has stopped growing and the likelihood of being able to obtain nuclear DNA for identification is much lower. However, even the hair itself (i.e. minus the root) will contain mitochondrial DNA, which can, and is, be used to determine maternal relationships.

It is sometimes possible to say whether a hair came from a child, because children's hair is finer than adult hair, although this is not always the case. For example, Napoleon had very fine hair as an adult. Bleached or dyed hair can be so identified and the period since the last bout of bleaching can often be gauged since hair usually grows at a rate of one centimetre a month.

Occasionally, damning evidence can be gleaned from animal hairs. In May, 1968, an 8-year-old boy, Christopher Sabey, was found murdered – strangled – in the village of Buckden in Huntingdonshire. After a very intensive investigation, during which four thousand statements were taken, suspicion fell on a young man named Richard Nilsson. It was then decided to take a further look at some seemingly trivial clues found on Christopher's body. They were three hairs, which looked as though they might have come from a dog.

Laboratory examination revealed them to be, indeed, dog hairs. They were white, tinged with fawn and could not have come from the Sabeys' black dog. Hairs found on Nilsson and hairs taken from his dog were of a similar colour. Neutron activation analysis showed that all three samples – from Christopher, from Nilsson and from Nilsson's dog – had very similar trace element characteristics. This was strong evidence, but it became stronger still. The police took hair samples from every dog in the village – all one hundred and forty-four of them – and sent them to the Home Office Forensic Science Laboratory in Nottingham. The process of collecting the evidence took three days and no comb could be used twice, for fear of contamination. Most were eliminated because they were the wrong colour, but fifty-four dogs did have the right colour. The trace element tests on these samples showed that the hairs on Christopher's body could have come from only three dogs, one of which was Nilsson's dog. Further evidence established Nilsson's guilt. At his trial, he was found guilty and sentenced to life imprisonment.

TEXTILE FIBRES

Allied to hairs are, of course, textile fibres. These are usually classified as being either natural or synthetic, although some fibres are manufactured from the cellulose derived from natural fibres and can, therefore, be regarded as being intermediate. Although there are many kinds of natural fibre, cotton is easily the most common. Nowadays, however, synthetic fibres are more widely used.

In the television adaptation of *The Holy Thief* by Ellis Peters, the mediaeval monk-sleuth Brother Cadfael was able to identify a murderer by showing that a gold-and-red fibre found at the scene matched those from the fine raiment of the aristocratic suspect. There was no poetic licence in this drama, since this kind of scenario often happens in forensic cases today.

Textiles are usually much easier to identify than hairs, because it is easier to maintain, or have access to, reference collections; although there are very many natural and synthetic fibres in use, these are finite in number and information about them is easily gained from the manufacturers. Also, textile fibres have characteristics that hairs do not possess, such as being dyed in all sorts of colours and woven in various patterns. Synthetic fibres have particular chemical compositions, which can be analysed in the laboratory.

Often, part of a garment is found at a scene and if it can be identified as being of a certain make; the date of its manufacture can be ascertained by simple inquiries to the manufacturer. In practice, this is more difficult than it sounds as today many garments and fabrics are mass produced, making it harder to pinpoint a single source as being the origin of a particular fibre. It is often of greater use to compare such a fragment with the garment of a suspected person, part of whose garment has been torn, in order to see whether the fragment came from that garment. This is, of course, a simple operation and it possible to say immediately whether or not the two parts fitted one another.

Two fibres may appear to have the same colour to the human eye, although the dyes used in each one may be different. In order to determine whether or not this is the case, organic solvents can be used to extract the dyes, which are then identified by the use of chromatography. However, such treatment would destroy the fibre as a piece of evidence and it often happens that there is very little material, perhaps just one short fibre, from the scene. In such cases microspectrophotometry is used.

A microspectrophotemeter is a microscope linked to a spectrophotometer, which is a device that measures the wavelengths of light. When light is transmitted through the fibre under the microscope, some wavelengths will be absorbed and others not. The pattern of wavelengths that pass through the fibre is recorded as a graph on the computer screen of the spectrophotometer. This spectrogram, as it is called, is essentially a description of the dyes in the fibre and can be compared with a database of spectrograms of a very wide range of fibres. This technique does not destroy the fibre.

The diameter, or width, of the fibre may also be an identifying attribute. This can be measured very easily under a microscope. In fact, colour and diameter are the two basic features that must be determined at the outset, since if these are compared with those of the suspect fibre and found to be different, there is little point in pursuing that line of evidence. On the other hand, if they do match, then there is every reason to go more deeply into the matter.

The next step is to establish the chemical composition of a fibre. In order to understand how this is done, it is necessary to say some words about the way synthetic fibres are made**. Fibres are made from chemical substances called polymers, which are very large molecules made of repeated, identical units known as monomers. So, if we designate a monomer as ∞ then a polymer would look like this:

∞∞∞∞∞∞∞∞∞∞∞∞∞∞∞∞∞∞∞∞∞∞∞∞∞

This picture is intended merely to illustrate the point; a real polymer is made up of thousands of monomers. Polymeric fibres are made by melting a polymer (either manufactured or natural, such as cellulose), or by dissolving it in an organic solvent. The polymer is then passed through the very fine holes of a spinneret, from which it emerges as numerous filaments, which harden into the fibres that are used to weave the cloth. The polymers have a regular arrangement along the long axis of the fibre. This regularity endows the fibre with the properties of a crystal and it is what imparts strength to it.

Unfortunately, from the point of view of the forensic scientist, polymers of this kind are not at all easy to dissolve, so, in order to discover the chemical composition of a fibre, it is not possible to subject it to gas chromatography in the normal way. The fibres must be heated to a very high temperature, decomposing the fibre into gases, which can be used in this technique. However, this technique is even more destructive than the one discussed above in relation to the chemical analysis of dyes. Happily, another technique, based on the optical properties of fibres, is available.

Light waves vibrate in all directions at right angles to the direction in which the light is travelling. However, when light passes through certain materials, including synthetic fibres, it emerges polarized, i.e. some of the vibrations will be at right angles and some will be parallel to the length of the fibre. This results in the light being refracted in two different ways. If a crystal of a kind that polarizes light is placed on this page, two images of the print beneath will be seen. The difference between the two refractive indices, which is usually very small, is known as the birefringence. Birefringence values of most synthetic fibres range from 0.001 to 0.18 and are specific to the various fibres, so they can be used to identify them. Thus, acrylic has refractive indices of 1.524 and 1.520, so the birefringence is 0.004; whereas Dacron, a kind of polyester, has a birefringence of 0.175.

A case investigated by the American forensic scientist, Richard Saferstein, shows how fibre evidence can be used in criminal detection. The body of a woman was found in an alleyway in East Harlem. She was identified as a local church member, who used to sell church literature in the area. Beside the body lay a large flower box and plastic liner, which were sent to Saferstein, together with the victim's clothing.

Light-brown wool fibres, navy blue wool fibres and red acrylic fibres were found in the box and the liner. These were compared with fibres from the victim's clothes, when it became clear that they all came from that source – light-brown overcoat, navy blue wool/polyester slacks and red acrylic sweater. However, additional fibres were found in the box and the liner. These were light blue nylon fibres and brown rabbit hairs. Significantly, fibres and hairs of the same kind were found on the deceased's overcoat, as well as some red nylon fibres.

This information was supplied to the police, who then learned that a man had recently sold a brown rabbit hair coat to a local man. They managed to obtain the coat from the purchaser and handed it over to Saferstein. The rabbit hairs from the coat were indistinguishable from those Saferstein had found in the box, the liner and on the victim's overcoat. The police felt they had sufficient evidence to obtain a warrant to search the suspect's apartment. There they found two rugs – one blue, the other red. Samples of these were sent to the laboratory, where examination revealed them to be made of nylon; and they were indistinguishable from those found on the victim's overcoat and, in the case of the light blue fibres, in the box and liner.

On the basis of this and other evidence, the man was arrested, put on trial, found guilty of murder in the second degree and was sentenced to life imprisonment.

TRACES OF BLOOD

One fact of which murderers seem totally unaware and about which, if they were aware of it, they can do nothing, is the fact that it is almost impossible to eradicate all traces of blood completely. Even after many years, very small traces of blood can be detected at the scene of a murder, if any blood had been shed at the time. Of course, burning a building in which a murder had been committed will destroy all such evidence, but, short of that, some bloodstain evidence will remain for a long time. Even smooth surfaces, such as bathroom tiles washed clean, will retain traces of blood.

At the scene of a brutal murder, large quantities of blood will be splashed all over the place. There will be drops and pools of blood on the floor, splashes on the walls, smears of blood here, smudges there, perhaps even a trail of blood leading somewhere else. All this has to be interpreted in an attempt to reconstruct events.

The interpretation of the pattern of bloodstaining is so dependent upon the experience of a forensic scientist that it is not possible to give more than a general idea of the kind of thing that is done. When a drop of blood falls to the floor, it will remain more or less spherical, if the floor surface is smooth, as in the case of linoleum or tiles. If the surface is rough or irregular, it will assume a star shape. However, care must be exercised when making interpretations of this sort, since a star shape will also form if the drop falls from a great height.

When a drop of blood strikes a wall at right angles to it, it will remain round in shape, but if it strikes the wall at, say, an angle of 20 degrees, it will form an elongated, oval shape. In fact, it is possible to determine the angle from which the drop came on the basis of its distortion, i.e. the extent to which it deviates from the circular form.

Blood falling at an angle, either on the floor or on a wall, will splash further on striking the surface, resulting in further stains beyond it. In other words, the smaller splashes lie on the side opposite to that from which the drop came. This can be shown diagramatically in the following way:

Direction of travel

These blood spatter patterns can be used to give the investigator an insight into the general nature of the crime that produced them; gunshot exit wounds commonly result in "high-velocity" spatter, whereas a beating may produce either high-or low-velocity spatter depending on the amount of force used.

Pools of blood may occur in two places at the scene, a fact that reveals that the victim moved, or was moved, from the place of initial attack to another place while still alive. The spurting of arterial blood over walls, and even ceilings, will reveal the place in which the victim was attacked, since the blood ceases to flow after death and the heart will not be able to pump it.

Some bloodstains can provide startling evidence. A sensational murder, the full story of which has not, even now, been fully revealed, occurred in the Bahamas during the Second World War, when the Duke of Windsor was governor of the islands. Sir Harry Oakes, probably the wealthiest baronet in the British Empire, was found bludgeoned to death in his bed. He was lying on his back. An attempt had been made to burn his body. In spite of the initial assumption that Oakes died in his bed, a tell-tale sign was discovered that suggested that his body had been moved. This sign became known as the blood trail that defied gravity, for blood from an injury above the baronet's left ear moved upward, over his cheek to his nose, then over the bridge of the nose and down to the other cheek!

After a very bloody murder, the murderer will sometimes try to conceal the body and clean the scene. Clothes may be washed, walls may be cleaned and floors may be scrubbed, but still a trace will remain. If is suspected that blood had been spilt in, say, a particular room, a minute investigation of it should reveal some traces of blood. Any stain or discoloration, however faint, that looks as though it might be blood, should be examined further. Often a microscopical examination will reveal some red blood cells adhering to an object, although, to the naked eye, it may appear quite free of blood.

When the presence of blood is suspected, some preliminary tests can be carried out. One of the best known of these is the Phenolpthalein or Peroxidase Test, of which there are several variations, the most widespread being the Kastle-Meyer Test. The suspect substance is removed by rubbing with a piece of tissue or filter paper and placed in a watch glass, whereupon 130 mg of phenolphthalein, 1.3 gm of potassium hydroxide and 100 ml of distilled water are added. This is boiled till clear; 20 gm of powdered zinc is added while boiling. A few drops of hydrogen peroxide is then added. If the solution turns pink, the result is positive. The test is sensitive even at dilutions of up to fifteen million.

Other methods of detecting blood have involved the use of certain sprays that either turn the stain a particular colour, or cause it to fluoresce. The spray luminol is used in dark conditions to detect blood on, say, a wall; if blood is present, fluorescence will occur. Leucomalachite green (which is, in fact, colourless) will appear as a green stain if blood is present. Spraying phenolphthalein mixed with orthotolidine will result in a pink colour in the presence of blood. However, a new reagent, Bluestar ®, which is based on luminol, is becoming more widely used as it is very sensitive to minute traces of both fresh and dried blood, and it does not require total darkness as it gives off a strong blue fluorescence that can be easily photographed with a normal camera.

Earlier, I used the word "preliminary" to describe these methods of blood detection. By this I meant that a positive result indicates that the substance present is most probably, but not certainly, blood. For example, the Kastle-Meyer Test will give a positive result if used on potatoes. Such tests are called presumptive tests by scientists. They serve to indicate that further work is necessary if a positive result is obtained; in the case of a negative result, however, one can conclude that blood is absent.

In order to confirm the presence of blood, a precipitin test is carried out. This is based on the immunological phenomenon in which antigens are formed in the blood of an animal when an antibody enters the system. Antigens react with antibodies, rendering them harmless. An antigen will react with only one kind of antibody, in other words, for each kind of antibody, a particular kind of antigen is needed. The way this is used as a means of identifying a substance as being human blood is to inject a rabbit, or other laboratory animal, with human blood. This causes the rabbit to form antigens in its own blood, with which to combat the foreign blood. A sample of blood is then taken from the rabbit and the blood serum isolated, which, because it contains antigens to human antibodies, is called human antiserum.

The antiserum is placed in a very fine capillary tube and an extract from the suspected bloodstain is added to it. If the extract contains human blood, a cloudy band will form at the area where the two liquids meet. If no reaction takes place, then the extract does not contain human blood, but it may well contain the blood of an animal. If this is suspected, the process can be repeated, using dog or cat blood (or blood from whatever animal is thought to have shed the blood) with which to inject the rabbit.

Nowadays, the test extract and the antiserum are usually placed in wells at opposite ends of a plate coated with agar, a gelatinous substance derived from certain species of seaweed. The two sera will move toward one another and a precipitate will form where they meet. A variation of this technique is the use of an electric current to make the two substances move faster toward one another.

LITTLE ITEMS OF EVIDENCE

It is obvious that so many little items of physical evidence can be used to reconstruct events; and space does not allow the consideration of all of them. Soil found on a shoe; a leaf stuck to clothes; paint smeared on a jacket; oil on a glove; all these things can link a person to a place. The principle is straightforward enough, the practice specialized and time-consuming. A specialist botanist may be needed to identify that suspicious leaf-fragment; a specialist geologist will have to be consulted about that strange reddish soil; and so on. There is, theoretically at least, no end to the kinds of clues one might find.

Tools often leave their "fingerprints" – the jemmy will make in impression on the wooden frame of a window, or a pair of scissors may have been used to cut a piece of paper, showing that it had not been torn. Great issues, as Sherlock Holmes might have said, may hang on such things. Much of the case for the murder of Julie Ward in Kenya was based upon the conclusive demonstration by Professor Austin Gresham that the bones had been cut with a heavy blade, not bitten through by a lion or a leopard, as was alleged by the Kenyan authorities.

In fact, the question of what the teeth of an animal can or cannot do may be very significant, as the following case will show. But before we discuss that case, let me say that the point of telling it lies not only in its interest as an example of a learned debate about whether an animal can or cannot do this or that thing with its teeth; its significance is far deeper.

Any forensic case is made up of little details. The task of the scientists involved is to try to reconstruct events; and if all the evidence seems to point one way, one has a scenario that is compelling. A story supported by many different lines of evidence will have a strength and integrity that is difficult to shake. However, one must be careful not to interpret the evidence in such a way that it is made to point one way only; often evidence can point many different ways. The probability of one's interpretation of the evidence – all of it – must be assessed. This is not always done, and I am bound to say that miscarriages of justice are known to happen as a result. The following case is an example of what can happen when bad forensic science, prejudice and confused thinking come together to perpetrate injustice.

A CAUTIONARY TALE

Michael and Lynne Chamberlain arrived at Ayer's Rock (Uluru) from their Queensland home on the evening of August 16, 1980. They set up camp east of the Rock. With them were their three children: Aidan, 6; Reagan, 4; and Azaria, a baby girl of nine and a half weeks. The Chamberlains were known to be people of good character. Mrs Chamberlain did not suffer from post-natal depression after the birth of her new baby; on the contrary, she was in very good spirits.

Shortly, before 8 o'clock the following evening, Mr and Mrs Chamberlain were at the barbecue area preparing a meal. Aidan and Azaria were with them; Reagan was asleep in the tent. After nursing Azaria, Mrs Chamberlain returned to the tent with Azaria and Aidan, where she put the baby to bed. Aidan then said that he was still hungry, so Mrs Chamberlain went to the car and fetched a tin of baked beans, then returned to the barbecue area with Aidan. Later, the time that elapsed between her leaving the barbecue area with Aidan and the baby and her return with Aidan and the tin of baked beans was estimated at being between five and ten minutes. The other campers noticed nothing odd about Mrs Chamberlain's manner on her return.

A short while later, Azaria was heard to cry – a cry that was heard both by Mr Chamberlain and other campers. Mrs Chamberlain starting walking back to the tent to attend to the baby. She suddenly stopped and, according to her own account, cried out: "That dog's got my baby!". Mrs West, one of the campers, later said that she heard the growl of a dog from the vicinity of the Chamberlains' tent and Mrs Chamberlain cry out: "My God, my God! A dingo has got my baby!". Mr West also said he heard the growl. Another camper, Mrs Lowe, said that the cry definitely came from the Chamberlains' tent and that it was the cry of a baby, not of a child. She also said that the cry was loud and sharp and that it seemed to stop abruptly.

Mrs Chamberlain said she saw a dingo (or dog) emerging from the tent, shaking its head. She could not see its snout properly, because it was below the light level and her view was obscured by shrubs and the railing that surrounded the camping area. She did not see the baby in its mouth, but guessed at the time that the animal might have had a shoe in its jaws. When she entered the tent, the baby was not in its bassinet. Mrs Chamberlain made a quick search of the tent, satisfying herself that Azaria had not simply fallen out of her cradle.

The alarm was raised and a frantic search began, but Azaria was not found. Drag marks were seen on the sandy ground and Aborigine trackers said that the marks were associated with dingo tracks. The police constable, who had arrived at the scene, and the park ranger, also saw the tracks. At length, the search was abandoned.

These, in brief, are the undisputed facts about what happened that night. At the trial, the prosecution asserted that, after going back with Azaria and Aidan to the tent, Mrs Chamberlain changed into a pair of track suit trousers, took the baby to the car, slit its throat, placed the body in a camera bag in front of the passenger seat, returned to the tent, removed the trousers, put on the clothes she had been wearing earlier, washed her hands (there was no water in the tent), then returned with Aidan and the tin of baked beans to the barbecue area, where she appeared perfectly calm and collected. These events were supposed to have taken place during a period of five to ten minutes. The prosecution could not suggest a motive for the murder.

Before we examine the evidence that was discussed at the trial, it is worth reading the testimony of the only person who was the nearest thing to an eyewitness to any alleged murder; Aidan, aged 6 years and 10 months at the time. This is what he said about the time in which Mrs Chamberlain was alleged to have committed the murder:

"After I finished my tea I said that I wanted to go to bed and mummy said that she would take me and Bubby [Azaria's nickname] up to bed. I went up to the tent with mummy and Bubby and I said to mummy is that all the tea I get. Mummy said that I could have some more tea. While we were in the tent mummy put Bubby down in the cot and then I went to the car with mummy and she got some baked beans and then I followed her down to the BBQ area. When we got to the BBQ area mummy opened the tin of bake beans and daddy said: Is that Bubby crying and mummy said I don't think so. Mummy went back to the tent and said: the dingo has got my baby. Mummy shouted has anybody got a torch and daddy went round and asked if anybody has got a torch. When mummy saw the dingo come out of the tent I was behind her but I didn't see the dingo come out of the tent."

It is true that this a child's evidence, but it strikes me as being remarkably honest. Aidan did not claim to have seen the dingo, a claim that would have supported his mother's evidence. It has been suggested that the boy may have been influenced by his parents, but anyone who has had any experience with children of that age knows that they will sooner or later blurt out the truth. It is not uncommon to hear a child say in the presence of adults: "I have nits in my hair," to the embarrassment of its parents, who may ask it not to say that in public. The child will not understand why it should not say what it knows to be true: "But I do have nits in my hair, mummy!".

The prosecution alleged that, after Azaria was killed, Mr and Mrs Chamberlain buried the body somewhere in the vicinity. Yet nobody saw either of them with digging implements or behaving oddly during that evening. They appeared shocked and distressed. There was a great deal of vegetational fragments and soil on Azaria's clothes when they were discovered about a week after the tragedy, four kilometres away from the Rock, near some dingo dens. A study of this evidence suggested that the body had been dragged through the various vegetational and soil types between the camping site and the place of discovery.

What was the physical evidence that was used to make a case against Mrs Chamberlain? First, there was the evidence of the blood; there was a certain amount of blood in the tent. The blankets covering Azaria were stained with blood, as were several other items. The track suit trousers that she was supposed to be wearing when she committed the alleged crime had some marks that resembled, but were not shown to be, blood stains; these were on the front and below the knee.

The Crown concluded that the blood on the trousers came from Azaria when she was killed in the car and that the blood in the tent came from Mrs Chamberlain's clothes when she returned there after the "murder". No evidence was presented to prove these assertions.

The prosecution alleged that there was blood in the car, especially under the dashboard and in the camera bag. However, it was shown that the forensic tests used to detect the blood were faulty and that what was said to have been blood was demonstrated not to be blood. However, a very small amount of genuine blood, inconsistent with what one might expect after a baby's throat had been slit, was detected. The source of this blood could have been a Mr Lenehan, an accident victim, whom the Chamberlains rescued in that same car. Also the Chamberlain children occasionally had nose-bleeds in the car. The amount of blood traces in the vehicle were what one might expect to result from the normal use of a family car. A Mrs Elston, who was given a lift by Mr Chamberlain in his car after the baby's disappearance, saw no blood. I should point out that the unreliability of the blood tests and the less-than-efficient manner in which they were carried out came to light only after the trial. At the trial, the jury believed that large amounts of specifically baby's blood was discovered in the car.

The prosecution's professional witnesses stated that the damage to Azaria's clothes was carried out with scissors and could not have been done by a dingo's teeth. However, the specialists in this field disagreed with one another and it eventually became clear that it could not be stated that a dingo could not have made the cuts in the clothing.

The prosecution alleged that the fact that the baby had been removed from its clothes was evidence of human involvement, since a dingo could not remove a body from clothes in such a manner as to damage them only slightly. However, experiments carried out at Adelaide Zoo, using a dead goat kid dressed in clothes, revealed that dingoes could remove clothes in this manner, albeit with somewhat greater damage to the clothes than was evident on Azaria's clothes. It is significant that the Park Ranger, Mr Roff, with his practical knowledge the habits of dingoes, maintained that a dingo would have no difficulty in removing a baby from its clothes in this way.

The prosecution alleged that dingoes are unlikely to attack people and kill children. In fact, frequent attacks on children and adults by dingoes were known from the Ayer's Rock area. Before the Azaria tragedy occurred, Mr Roff had written to his superiors about the dangers to people from dingoes, writing that "children and babies can be considered possible prey". Shortly before the Chamberlains visited the Rock, a dingo had removed a pillow from beneath the head of a camper.***

In short, there was no evidence against Mrs Chamberlain. Much of the forensic evidence was discredited. But the jury at the trial, probably unduly influenced by the blood evidence, returned a verdict of "Guilty".

On October 29, 1982, Alice Lynne Chamberlain was convicted on a charge of murder by the Supreme Court of the Northern Territory. Her husband, Michael Leigh Chamberlain, was convicted of being an accessory after the fact. The convictions were subsequently upheld by the High Court. Both Mr and Mrs Chamberlain persistently denied the charges made against them. After five years in prison, Mrs Chamberlain was released, when a Royal Commission of Inquiry concluded that there were "serious doubts and questions as to the Chamberlains' guilt".

In his report, the Royal Commissioner, Mr Justice Morling, commented on the forensic evidence adduced at the trial:

"The question may well be asked how it came about that the evidence at the trial differed in such important respects from the evidence before the Commission. I am unable to state with certainty why this was so. However, with the benefit of hindsight it can be seen that some experts who gave evidence at the trial were over-confident of their ability to form reliable opinions on matters that lay on the outer margins of their fields of expertise. Some of their opinions were based on unreliable or inadequate data. It was not until more research work had been done after the trial that some of these opinions were found to be of doubtful validity or wrong. Other evidence was given at the trial by experts who did not have the experience, facilities or resources necessary to enable them to express reliable opinions on some of the novel and complex scientific issues which arose for consideration. It was necessary for much more research to be done on these matters to determine whether the opinions expressed at the trial were open to doubt."

Yes, the evidence was presented in such a fashion as to suggest guilt at every turn. It was not simply a matter of the need for "more research"; what was already known should have suggested caution in interpreting the evidence. It is the fact that every piece of evidence was interpreted by the prosecution in the most damning, if unconvincing, way that is the real cause for concern. Forensic specialists, as we saw in an earlier chapter, are sometimes apt to accommodate lawyers in these matters.

But the story of Lynne Chamberlain tells us a great deal more about the nature of evidence. This innocent woman, having lost her daughter in dreadful circumstances, was accused of her murder on the basis of – what? Nothing whatsoever. There was no evidence against her, no *prima facie* case to answer at all. The case was not a case. Why, then, did all this happen?

The Chamberlains were Seventh Day Adventists. This made them unpopular with certain sections of the community. There was certainly a great deal of prejudice against them. There was readiness to believe the worst about them. It was even suggested, needless to say in the absence of any evidence, that Azaria was ritually slaughtered.

But this is not the worst of it. The case against Mrs Chamberlain was based entirely on the belief that the dingo story was false. If there was no dingo, then Mrs Chamberlain was guilty, in spite of the fact that she had no motive (quite the opposite, one would have thought) and she had no opportunity, since what she was supposed to have done, in the company of her son, in five to ten minutes, is simply not credible.

Let us suppose, for the sake of the argument, that Azaria had not been taken by a dingo. Why does it follow that Mrs Chamberlain murdered her? In fact, it does not follow; it is a complete non sequitur. Yet the forces of law and order managed to make a case against her. Why was it never suggested that someone else could have committed the murder? Surely this is the most likely explanation, if it could be shown that the dingo did not exist. Yet this suggestion was never made. Even after Mrs Chamberlain's release this question continued to be resisted. At the end of his report, the Royal Commissioner wrote:

> "In reaching the conclusion that there is a reasonable doubt as to the Chamberlains' guilt I have found it unnecessary to consider the possibility of human intervention (other than by the Chamberlains) in the time between Azaria's disappearance and the finding of her clothes. It is difficult, but not impossible, to imagine circumstances in which such intervention could have occurred. It is not inconceivable that an owner of a domestic dog intervened to cover-up its involvement in the tragedy or that some tourist, acting irrationally, interfered with the clothes before they were later discovered by others. There is not the slightest evidence to support either of these hypotheses but the possibility of human intervention is another factor which must be taken into account in considering whether the evidence establishes the Chamberlains' guilt beyond reasonable doubt."

"*Unnecessary to consider the possibility of human intervention.*" Why? Only intervention by the Chamberlains can be considered. Why? "*There is not the slightest evidence to support either of these hypotheses.*" What evidence was there to support the accusation against Mrs Chamberlain? "*It is difficult … to imagine circumstances in which such intervention could have occurred.*" Is it, then, so easy to imagine that a mother would slit the throat of her baby? There is no rational assessment of probabilities in all this, still less is there any understanding of human nature. The Commission concluded that there was only "a reasonable doubt" as to the Chamberlains' guilt, although two pages later this was changed to "serious doubts". It is hardly necessary to comment further.

---

* More correctly, the refractive index is the ratio of the speed of light through a vacuum (rather than through air) and its speed through a medium. For our purposes, however, we can take the speed of light through air as our standard.

** Natural fibres are often identifiable on the basis of their structure, much like hairs, although the other techniques discussed in this chapter may be used.

*** Since the Chamberlain case, several attacks by dingoes on adults have been recorded.

CHAPTER TEN

THE CRIMINAL MIND

*The mind is its own place, and in itself
Can make a heaven of hell, a hell of heaven*
John Milton
Paradise Lost

I was sitting in a court-room, looking at the man in the dock. He was an extremely violent man, who had killed or severely injured a number of people in a particularly unpleasant manner. I had already given my evidence and been cross-examined; I remained in the courtroom to hear the rest of the trial. The man looked back at me from time to time, narrowing his eyes, since my evidence seemed to incriminate him. His whole bearing was a threat. He was found guilty of manslaughter, not murder, and the judge sentenced him to four years imprisonment. Later, the young pathologist who had worked on the case told me, in a jocular kind of way, that he expected the fellow to turn up at my house after his release to wreak his revenge. I did not find this comment as amusing as he did.

Why do some people commit crimes? Why do they do these things, while most of us seem to have no difficulty in living law-abiding lives? What turns a human being into a monster?

In this chapter we will try to find answers to these questions. They are questions that have to do with the human mind, a subject fraught with difficulties. If some of my comments seem strident, I crave your indulgence, for these are matters of supreme importance and cannot be treated in a superficial manner. Nettles will have to be grasped firmly; lions bearded in their dens and problems looked at squarely in the eye. Many people have an almost blind belief in science and they often find themselves unwilling to contradict its findings; but science is not omnipotent and it frequently fails very badly, especially in the sensitive area of human behaviour. Not all science is scientific, as we shall see.

However, before we proceed I would like to make one point absolutely clear. Some academics are notorious for their tendency for bitter disputation, concealing personal antipathy with an affectation of scientific debate. I assure the reader that the criticisms I make below are not derived from any such base motive; and for this reason, I will not name any individuals with whom I disagree. I am not trying to be clever or amusing when I say that some of those people with whom I most strongly disagree about such matters are among my closest friends.

FALSE AND ILLOGICAL REASONING

There is a story, which I sincerely hope is apocryphal, about a scientist who studied frogs. He would place a frog upon the laboratory bench, then, moving close to the animal, he would shout loudly at it. Invariably, the frog would jump away along the bench. The scientist then decided to take his investigation further. He took a frog and cut off its legs, then placed it on the bench as before. Moving closer, he shouted loudly, but the frog did not move. Why? Clearly, the frog did not jump, because it could no longer hear; and it could no longer hear, because its legs had been removed. Conclusion: frogs hear with their legs.

This is the image of the idiot savant; the learned man who, for all his erudition, is a fool. While I have never met with such an appalling example of false reasoning from a scientist, it is nevertheless true that some areas of science abound with examples of totally illogical reasoning that are not worlds removed from our amusing story of the unfortunate frog.

Psychology, by it very nature, is a subject that should interest anyone interested in people and society generally. However, I see it as being largely a descriptive, rather than an explanatory or predictive, field of human inquiry. So, before we look into the ways psychologists and others have tried to explain or predict human behaviour – criminal or otherwise – we will first consider its contribution to the description of criminal behaviour.

Psychological, or "offender", profiling is the field of investigation concerned with describing the mental outlook and general background of a person who has committed a crime, with the aim of assisting the police to find him. Clues left behind by the criminal can be used to build a picture of his mind. For example, it is said that most serial killers are white, not black, a comment that would be regarded as racist if it had been the other way round. If a crime took some time to execute, such as the murder, rape and subsequent mutilation of the victim, it is construed that the criminal must have been familiar with the area, since he would not have spent such a long time at a place unfamiliar to him. It is also considered likely that a serial killer, especially one whose crimes are particularly horrific, would have had an unhappy childhood, or have come from a broken home, or had been physically abused as a child.

So far, so good; it all makes sense. More to the point, such profiles are almost always shown to be true when the offender is caught. So, is this the end of the matter?

Well, no, for there is a great deal that is worrying about how such information is used and interpreted. Consider first this question: "Is it not intuitively likely that a violent criminal would be unhappy?". The answer must surely be "Yes"; and one is tempted to say that one did not need a psychologist to tell us this. Worse still the problem generated by asking the following question: "If a criminal has X characteristics, does this mean that people with X characteristics are criminals?". I realize that this is not what offender profiling specialists are saying, but it is what other psychologists seem to be saying, for, as a group, psychologists try to explain human behaviour, in other words, they ask the question the other way round – the second half of the question in the last sentence.

We can all agree that individual human beings do have distinctive personality traits, but it is almost impossible to say how they acquired them in the first place. We may be tempted to say that so-and-so reminds us of his father because he has the same morose temperament and that, presumably, this personality trait is genetic. On the other hand, it might be that he is like his father because he grew up in the gloomy atmosphere generated by his father's morose temperament and that, therefore, the origin is environmental.

Human beings are such complex creatures that I believe that their behaviour is simply not amenable to study. (Of course, it is quite possible to study the physical traits of people.) I think it is true that one can only get to know people; studying them, in the sense of being able to measure and predict their behaviour on the basis of scientific principles, does not seem to me to be possible. The human attributes that most concern us are simply not measurable.

Let us say that a man is six feet tall and that his son is three feet tall. There is no doubt that the man is twice as tall as the boy. But what would you be likely to answer if I asked you how happy you are? Or if I asked you how nice your friend is? Or how much you would like to have a million pounds? These are very real questions to most people, but the qualities they are concerned with are not measurable or quantifiable. You cannot say that you are five feet happy or that your friend is ten kilograms nice or that the extent of your desire to have a million pounds can be measured in gallons.

Nevertheless, let us try to be scientific and make an attempt at measuring these qualities. Take the quality of niceness. Let us say that I give you a list of people and I ask you to grade them, in increasing order of niceness, along a scale of one to ten; the nicest will score ten, the least nice one. You arrange the people in that order and hand me the results. Will they be a true measure of this quality?

No, they will not, for three reasons. First, the scale of one to ten that I gave you is only superficially scientific, because it is not a true, metric scale. In other words, the person who scored six is not twice as nice as the person who scored three, unlike the example of the height of the man and the boy. The person who scored ten is not ten times as nice as the person who scored one, and so forth. There are no objective units (feet, kilograms, gallons) to the "measurement", so it is not really a measurement at all; there are no units (feet, pounds, hours) of niceness.

Secondly, if I had graded the same people when you were doing your own grading, we would almost certainly have found that our results differed; I would have scored some people higher and others lower than did you. Thirdly, your grading would have depended very largely on how you were treated by these people and the extent of your knowledge of their characters. If I asked you to grade them again after six months, you would probably arrange them differently, whether or not you remember the results of your first attempt. It may be argued that the same applies to the heights of the man and the boy, since the boy will have grown taller in that period and the man may not. However, if we both measure the man and the boy again, we will, again, get the same results as one another, assuming we both know how to use a tape measure. The results of the heights are objective, of the niceness subjective, yet we both know exactly what we mean when we refer to someone as being nice, even though I might find a particular person very nice and you might find him extremely unpleasant.

Attempts to measure human attributes in this way are both meaningless and harmful. Senior managers in business and industry and even, astonishingly, the staff of universities and hospitals are now being graded in this manner. Forms are circulated by managers to others down the hierarchy, asking them to score the ability of staff members in this way, usually on a scale of one to five. According to this system, the intelligence, efficiency, ability to work hard, friendliness, endurance and other human qualities of individuals are directly measurable and, worse, capable of meaningful comparison with other individuals. Certainly, we may validly form judgements about abilities of people on the basis of our knowledge of them, but to pretend that the figures produced by the scales of psychometricians have any objective validity or consistent nature is a complete fallacy.

Now we come to the most dangerous aspect of behavioural studies – the fashionable deterministic theories that assert that we are what we are and cannot help it. Some such studies come within the realm of science, but only in the sense that their practitioners call themselves scientists. Other studies fall into the field of sociology, while some are a mixture of the two.

SOCIOBIOLOGY

We will begin by considering the contributions made by the field known as sociobiology. This area of biology seeks to explain human social behaviour in terms of animal behaviour and biology; its basic tenet is that all behaviour, including social behaviour, is genetically determined.

As far as human society is concerned, what sociobiologists are saying is this: human beings exhibit certain character traits; they organize themselves in groups (families, tribes, nations, etc.); they go to war; they co-operate; they care for their young; and so forth. These traits, they say, are controlled by the genes and human nature, therefore, evolved by means of natural selection, in much the same way as the giraffe's neck or the elephant's trunk.

In short, it is asserted that human nature is genetically predetermined, in the sense that human genes do not merely allow people to do certain things, they compel them to do so. One of the arguments used to justify this belief is the fact that certain habits of behaviour seem to be very widespread among people all over the world. For example, they say that, since war has been such a common human activity among all peoples throughout history, it must be a genetic trait. The author of a recent book justified this kind of belief saying that, since morality is older than the church, culture older than Babylon, society older than Greece, trade older than the state, these traits must be genetic ones and that their roots are in the "missing links" with other primates.

The first flaw in this argument is that, simply because a trait is widespread, it does not necessarily follow that it is genetic. Effectively all the inhabitants of France speak French, but this does not mean that there is a French language gene. Although the genetic make-up of the French people allows the speaking of French, it does not cause it, since a French child brought up in England will speak English, not French, and an English child brought up in France will speak French, not English. Similarly, anyone can use a computer if taught how to use one, yet it must be clear that no gene for the use of computers ever evolved.

These points can be made more comprehensible by considering the following analogy. One can study the structure of a car in great detail. One can be intimately acquainted with its every component and understand its mode of function precisely. But, even if one were to be familiar with every last molecule of the car, one would not be able to predict where it will go tomorrow, for that depends on the person who drives it; in other words, on an external factor that does not come within the reach of the studied elements. And so it is with behaviour. The hardware of our brains may be the consequence of genetics, but how we use them is largely dependent upon the external factors of our environment, including everything outside ourselves.

The second fallacy is that the fact that a behavioural characteristic is older than the name given to it or the establishment of a formal institution dealing with it does not mean that it is in the genes. Morality being older than the church or any form of institutionalized religion does not mean that morality is the product of natural selection. As for the notion that the roots of such behaviour lie in the so-called missing links between humans and other primates, this is an assertion unsupported by any evidence; it is not in any sense a scientific statement, but an arbitrarily adopted opinion.

Nevertheless, despite these objections, for many years sociobiologists have persisted in claiming that human nature is encoded in the genes. This view is impossible to reconcile with the known facts of genetics, since there is no known genetic technique that can detect the presence of behavioural genes that are common to everyone. (The genetic ideas of sociobiologists, it has to be made clear, are not based on any practical scientific investigation.) If everybody is genetically predisposed to wage war or to be moral, then such a fact cannot be found out. On the other hand, if such genetic traits are present in some people, but not in others, then there is no such thing as a genetically uniform human nature, as the sociobiologists claim.

Sociobiologists try to explain the evolution of certain phenomena, such as aggression, slavery or the caste system, by invoking their presence among animals. For example, slavery in humans evolved from animal origins, for do not ants indulge in slavery? The blunt answer is, no, they do not. It is true that such things called slave-making ants exist, but their existence proves nothing. Slavery is a human term that describes the exploitation of one human being by another. Ant "slavery" is concerned with the exploitation of one species by another, since no ant enslaves a member of its own species. Slavery in ants is thus more akin to the keeping of domestic animals by humans. What sociobiologists have done is to take the word "slavery", misapply it to ants and then "explain" that the very different human institution evolved from animal origins.

It is in the field of morality that sociobiology has caused most confusion and damage. According to sociobiological theories, morality, or altruism, is a genetically inherited trait. The claim is that altruism evolved because it allowed individuals to enhance the chances of their genes being passed on to future generations. This claim is dependent on another claim, namely, that organisms, including human beings, are vehicles for the propagation of genes. In other words, the organism is merely the tool that the gene uses to make more genes like itself. This notion, popularized by the term "selfish gene", is yet another one of the unsupported claims of sociobiologists. This theory says that people are nice to one another because they stand to benefit genetically. All sorts of hypothetical examples are invented to support this notion, but no actual examples from the real world are ever put forward. For example, sociobiologists say that if I help you on one occasion, this would increase the chances of your helping me on a future occasion. This may well happen, of course, but it equally well may not. If you help a blind man to cross the road, how would such an act benefit you in the future, especially since you are very unlikely ever to see the man again?

Such real situations are never addressed. Sociobiologists concentrate on such questions as the assistance given to the rearing of the young by individuals who are not themselves parents. The example of social insects, such as wasps and bees, is frequently used as an illustration. Worker bees, who are sterile females, help in the rearing of the Queen's offspring, apparently because they share some genes with them. No evidence for this claim is given, however. It is merely a clever idea, but genuine science needs more than clever ideas. People are known to give their lives for total strangers or even an idea or a belief. In other words, they do not stand to benefit genetically from their acts of self-sacrifice, but these real examples are countered by the claim that such people are still acting out of self-interest, since to sacrifice themselves is what they want to do! However, this cannot increase the chances of these people's genes surviving, unless they are beyond the age of reproduction. In any case, we know that younger people may behave in this way. In the end, self-interest is what sociobiologists say is self-interest, and the concept loses all meaning.

One sociobiologist wrote in one part of his book that altruists are "very, very few", but in another part of the book he says that "Our minds have been built by selfish genes, but they have been built to be social, trustworthy and cooperative". Both statements cannot be true (in any case, the second one contradicts itself), but they are the natural consequence of trying to maintain a genetic theory at the same time as maintaining a liberal, humane outlook. Sociobiologists are usually kindly, well-meaning souls, but they are monumentally confused. Further evidence of this statement is the fact that many of them do not believe that the selfish gene should dictate morality. One of the most famous exponents of the subject once wrote that we must "rebel against the tyranny of the selfish replicators". This is a very interesting remark, for two reasons. First, it undermines the whole basis of sociobiology, since it prompts the question, "Why?". Why should we resist the tyranny of the selfish gene, if that is the only reality? Secondly, it is intellectually dishonest, since what started as an interesting, albeit unconvincing, idea has ended up as a fact to be "rebelled against".

This brings me to the most fundamental objection to sociobiological theory. If genetics controls our behaviour – our actions, our thoughts, our beliefs and our pronouncements – then it also controls the pronouncements and beliefs of sociobiologists. If we, poor ordinary human beings, cannot help doing or thinking or saying the things we do, think or say, then nor can sociobiologists. It follows that there is no particular reason for believing what the sociobiologists say, since they are pre-programmed to say them. If behaviour is, indeed, controlled by the genes – if we have to say what we say and have no control over the matter – then no opinion, sociobiological or otherwise, can possibly have any value. There is no point in the sociobiologists preaching to us about genetics, since by so doing they saw off the branch they are sitting on.

I have confronted sociobiologists many times with this argument. Their response, incredibly, is to say that we are, in fact, in control of our actions, because we have free will. This is quite an astonishing response since it completely undermines their whole subject. If free will controls our actions, then what on Earth is the whole subject about?

GENETIC PREDICTORS OF CRIMINALITY

Is there, then, no genetic basis for criminality? Before we try to answer this question, let us look into the genetic researches that have been carried out. During the nineteenth century, several attempts were made to predict whether or not a person had criminal tendencies. Thus, assassins were thought to have well-developed jaws and thick, dark hair; rapists had short hands and narrow foreheads; thieves had oddly shaped heads; swindlers were heavy and pale; and so forth.

While it is easy for us to smile patronizingly at such simplistic theories, many (I do not say all) of the modern ideas that supplanted them are no less simplistic, notwithstanding their sophisticated and scientific veneer.

The ideas in question are also concerned with biological differences, albeit differences of a "deeper" significance. Instead of looking at narrow foreheads or jutting jaws, defects of the nervous or endocrine systems are invoked as causes, or predictors, of crime. Thus it has been claimed that glandular dysfunction is correlated with aggressive behaviour, although there is hardly any evidence to support this assertion.

Other physical conditions have been invoked as explanations of criminal behaviour. Senility, vitamin deficiencies and certain diseases may cause people to behave irrationally, but not necessarily in a criminal manner. Abnormal electroencephalograms (EEGs, or "brain-waves") have been found to occur in many murderers. In one study, no less than fourteen out of eighteen people who had committed apparently motiveless murders had abnormal EEGs.

However, the interpretation of these results presents a difficulty: do abnormal EEGs and glandular dysfunction cause criminal behaviour, or are they the consequences of it? We have already discussed the dangers of correlations as causes and effects in Chapter 2. Do criminals' EEGs undergo a change when they have committed a crime and when they are being held in custody? It seems quite plausible that they do, although we do not know – the researches have not been carried out.

In order to circumvent the problem of what is cause and what is effect, the search for genetic predictors of criminality moved into the subject matter of genetics itself – the actual chromosomes that bear the genes. It was argued, quite reasonably, that the genes, being present from the very beginning of an individual's life, can only be causes, not effects.

Men and women differ genetically, because men have two different kinds of sex chromosome (one of each), while women have two of the same kind. The kind that both sexes have is called the X chromosome; the one present only in men is called the Y chromosome. So, men can be designated XY and women XX. Now, it so happens that some men have an additional sex chromosome; sometimes the complement is XYY, sometimes it is XXY.

Researches in Scotland have shown that a large number of mentally subnormal men had chromosomal abnormalities of this sort, especially the XYY state. Further studies in the USA went a step further, suggesting that XYY men were more disposed to act in a criminal manner. Yet again, further British research appeared to show that the opposite was true, that XYY men were less aggressive than other criminals who had the normal XY complement.

The problem with all these findings is not so much that they seemed to contradict one another, but because they did not take into account the percentage of XYY men in the population as a whole. In other words, if, say, ten per cent of the male criminal population are XYY men, could this not be simply a reflection of the possibility that ten per cent of the male population as a whole (criminal and non-criminal) were also XYY? This very important aspect was never looked into, which means that the results simply cannot be interpreted meaningfully.

Studies on identical twins separated at birth and brought up by different families, and studies on adopted children, have been conducted with the aim of establishing whether criminality was inherited or acquired. Briefly, the findings of these studies were inconclusive, largely due to the fact that many methodological problems confused the issue to such an extent that interpretation became impossible. For example, some studies investigated very small numbers of individuals. In the case of twin studies, the adoptive parents were of a background similar to that of the natural parents, making it difficult to conclude what the results really meant. Thus, the children of criminal parents, when brought up by another family, did sometimes show a tendency toward criminality, but it was not clear whether this derived from the general social background of the adoptive parents or from the children's genetic inheritance.

So, what can one conclude? Is criminality inherited in the genes, or is it acquired from experience? Or is it a mixture of both, nature and nurture each doing its bit? The answer to the last question must be "Yes", yet, while it would not be fair to say that it is akin to sitting on the fence, it is a position that tends to stifle any further discussion. Theodore Bundy, himself a lawyer, a psychologist and a serial killer, once said: "If anyone's looking for pat answers, forget it. If there were, the psychiatrists would have cleared this up years ago". A clever statement from an intelligent, albeit criminal, man. But it is not, in fact, as clever as it sounds, for it jumps the gun. It assumes that the solution lies in psychology, which it may not.

THE SUPPOSED ROOTS OF CRIMINALITY

Where, then, does the answer lie? Many people believe that it does, in fact, lie in genetics, or at least, in biology. This is a widespread opinion, held by many able scientists, as well as other intelligent men and women. It is only fair that I should mention this fact, because I disagree with them. The reader should understand, therefore, that the views that I am about to put forward are not necessarily the orthodox views of science as a whole. There are two fundamental reasons why I differ.

The first reason is that there is no evidence that criminality is genetically inherited, nor can any such evidence ever be found. If this seems a startlingly extreme point of view, consider what we mean when we use the word "criminality". I think everyone would agree that criminality is a kind of behaviour that offends our moral beliefs, or behaviour that is not allowed under the law. Therefore, it is a moral concept, or a legal one, but it is not a scientific one. No amount of genetic study will tell us whether a certain kind of behaviour is "right" or "wrong", legally or morally speaking. The students who opposed the tanks in Tiananmen Square were criminals in China, but were heroes to the rest of the world.

Am I, then, suggesting that all scientific research into the causes of criminality should be abandoned? Not quite, but we must be absolutely clear about what traits we are investigating. Aggression has been equated with criminality, but an aggressive person is not necessarily a criminal and may, indeed, be a worthy member of the community. It all depends on the channelling of those aggressive "instincts", inherited or otherwise. Scientists, artists, explorers, athletes, mountaineers, business managers and many others are often very aggressive people, but their aggression is directed toward doing something useful. Therefore, the studies that claim that aggression is heritable (and such a claim, so far, has very little to support it) are making the logical error of saying that criminality is also heritable, simply because they have linked the phenomena of aggression and criminality in their minds. By all means let us continue to study the scientific question of whether aggression is heritable or not, but let us not pretend we are investigating criminality when so doing.

Another, somewhat more disturbing, point needs to be raised. Geneticists who claim to be studying criminality are almost always concerned with violent crime, hardly ever with what we might call "white-collar" crime. This may be because, while it is easy (although erroneous) to equate aggression with criminality, there is no such conveniently overt kind of behaviour associated with defrauding one's employer, or betraying one's country, or even slipping some cyanide into someone's drink. If "criminality" is heritable, what convenient handle can we use to claim that there is a gene for it? The answer, of course, is that there isn't one.

We come now to the second reason why I do not believe that genetic or biological theories can adequately explain the causes of criminality. Crime rates rise and fall at different times. What are we to conclude from this? Is it likely that the genetic make-up of people changes back and forth over the years to produce such fluctuations in criminal behaviour? The answer must be "No", especially since the time periods in which these changes take place are often much less than the period of a single generation. To invoke genetics as a cause of crime, therefore, is to ignore this fact and the fact that crime rates do rise and fall in association with changes in the economic or political environment. This does not mean that we are necessarily dealing with a case of cause and effect, but it is a correlation worth thinking about and exploring further.

It is not only some scientists who hold deterministic opinions about the causes of crime. Academics at the other extreme end of the spectrum try to advance the exact opposite explanation of human behaviour. According to these cultural determinists, everything a human being does is the result of their environment and upbringing. They claim that all ideas and beliefs are culturally determined. Where right and wrong were in the genes with the sociobiologists, they are in the culture with the cultural determinists. But it will not wash, for much the same reasons. If all beliefs are predetermined by one's culture – if one cannot help believing certain things, because of the way one is brought up – one is entitled to ask whether the same is true of the beliefs of the cultural determinists themselves, since they, too, are products of specific cultures. So, according to their own theories, we have no particular reason to believe them. Like the sociobiologists, they have painted themselves into a corner. Both schools of thought would have us believe, in the teeth of the evidence, that behaviour and morality are thrust upon us against our will.

The truth must be that both nature and nurture play a part in all this, but not in the simplistic way sometimes put forward. It is not simply a matter of nature (genetics and physiology) contributing X per cent and nurture (environment and culture) Y per cent. Both work on one another in very complex ways and the resulting behaviour of a human being cannot be dissected in such a way as to allocate some expressions of it to nature and others to nurture.

So, do we now have the answer? Unfortunately not. People have known about nature and nurture for millennia, but that has not helped humanity to find the answer. The difficulty, I think, lies partly in the fact that many answers have been very confidently put forward by different authorities; those authorities do not only disagree with one another, but they often refuse to see that the others even have a point of view.

Geneticists may tell you that criminality is the DNA; psychologists that it is in the upbringing; sociologists that it is in the culture. There is an element of truth in all of them, but their answers, individually, fail to satisfy. When the geneticist talks of DNA, we wonder why our friend, who has such a short fuse, is nevertheless the best person we have ever met. When the psychologist talks about unhappy childhood, we wonder why so many people who have had such childhoods manage to lead normal, useful lives. When the sociologist talks of cultural indoctrination, we wonder why so many people rebel against the culture of which they are a part.

Harold Shipman, the serial killer of old women, is said to have committed his crimes because he had an unhappy childhood and because his mother died when he was young. Why, then, are other people able to refrain from murder, even though they had similar experiences? It seems to me that psychological answers to these questions are wisdom after the event. The argument seems to run as follows: Howard Shipman was a serial killer; Howard Shipman was unhappy as a child; therefore, childhood unhappiness was the cause.

Note that no reason is giving for selecting that particular "cause"; any number of other "causes" could have been invoked. Childhood unhappiness is chosen, one has to conclude, because it is the subject that a particular psychologist is interested in. The truth is that we do not know why Shipman did what he did – guessing is not going to give us the answer in his case, nor will it help us to deal with the problem of criminality generally.

I do not mean to say that psychological explanations are never the cause of a particular crime. Peter Sutcliffe, the Yorkshire Ripper, is clearly mentally unhinged, to the extent of having such delusions that would make him produce paintings of himself lying dead at the foot of the glorious revolution.

Nora Tierney, who murdered 3-year-old Marion Ward for no obvious reason, is another case in point. The crimes of such people seem genuinely to be caused by a psychological imbalance so severe that they cannot be considered to be of sound mental health and may well be beyond help. So, while psychology can explain certain crimes, it cannot explain crime as a phenomenon.

THE CAUSES OF CRIME

What, then, are the true causes of crime? The pressing need for an answer to this supremely important question is matched by the great difficulty of finding such an answer. The possible underlying causes – physical, psychological, moral, social and economic – have been examined and have failed to provide a satisfying explanation. This is puzzling, since, theoretically, the answer must lie in the way in which one or more of these potential sets of causes affect our behaviour. We feel that the answer should be within reach, yet it continues to elude us.

Why? The reason that is usually given is that the causes of crime are too complex and many-stranded; it is simply not possible to come up with an easy answer. This is undoubtedly true, but it is not only easy answers that are elusive; complicated answers are equally difficult to find. I think this is partly due to the fact that answers or solutions are usually proposed by individuals or groups starting from a narrow base. If a politician proposes an answer, he will probably be strongly affected by his own political or ideological position. Academics studying the problem will often be pursuing a particular scholarly aspect that interests them. Others, such as churchmen, social workers or police officers, may be too close to the problem to be able to stand back and take an objective look at it and may look at the problem from their own restricted point of view. Hence, ideological, scientific and emotional positions often dictate specific answers, which are often unconsciously preconceived, with the result that it is very difficult to get a complete picture. Also, because of the political and emotional nature of the subject, genuine misunderstandings often occur and sometimes deliberate obfuscations are used.

An example of the sort of confusion that is a mixture of genuine misunderstanding and politically induced manipulation of the truth can be seen in the sterile and often bitter debate between right-wing and left-wing sympathizers about the role of poverty as a cause of the rise in crime. Left-wingers insist that poverty is a main cause of crime and that crime would be reduced if more is done for the poor. Right-wingers retort by saying that most poor people remain law-abiding and that it is an insult to the poor to suggest that they are less law-abiding than other citizens. Every so often an item appears in the newspapers, saying that some well-known figure or organization has supported one side or the other and that this is seen as a victory for that side's point of view. The debate has been pursued by the two political opposites for many years, without any progress.

This is not surprising, since both sides are not only ideologically wedded to their own explanation, but, more importantly, they are at cross purposes with one another without, apparently, realizing it. The left-wingers say that people are more likely to break the law when they are poor. It is the poor man who will steal to feed his family. If he were not poor, he would not steal, since he would be able to acquire what he needs without having to break the law. This seems reasonable enough. The right-wingers, on the other hand, say that most poor people do not steal, even though they cannot afford to acquire what they need or want. Being poor, they say, does not mean that it is right to steal. This, also, is true. In short, the left says that poverty causes crimes and the right says that poverty does not excuse crime. Both are right, but about different things. Either they cannot, or will not, see this.

It is this kind of public debate about the causes of crime that confuses the issue and hinders progress. In reality, such debates are not concerned with solving the problem; the problem seems to be used as an ideological weapon by certain groups with axes to grind. The same applies to public debates in which the role of the police in fighting crime is discussed; two sides, which we can call, for the sake of convenience, left and right, emerge and argue about the efficacy and the morality of the police. The discussion turns into an argument in which the police are used as a political stick with which to beat the other side. One side will talk about police brutality, the other about the need for more bobbies on the beat. Once again, the debate is diverted into matters that are quite off the point.

These days it seems that almost everyone wants a public debate about something. The genesis of criminal behaviour is a valid and urgent subject for such a debate, but it will get nowhere if it continues to be treated as a political issue rather than a human one. (I have used the word "human" deliberately, shying away from the word "social", because of its political connotations in this context.) To begin this debate I propose to take a fresh and, as far as possible, objective look at all the possible causes that could explain the existence and the increase in crime; for crime, at least violent crime in Britain, doubled in the last twenty years or so of the twentieth century and it continues to rise.

The reader must not expect any profound or illuminating insights from what follows. What I will do is set down the problems as I perceive them and see whether certain causes reveal themselves as being more likely explanations than others. I present these thoughts for consideration, not as answers based on irrefutable evidence. Evidence will be needed to assess the validity of these ideas.

We have already considered the question of inherent biological make-up as a cause of crime and concluded that there are grave dangers in the conclusions that have been proposed by some criminologists. This does not mean that a person's physical state cannot affect their behaviour in a way that could be called criminal. A man tormented beyond endurance may well become violent, even though he would not behave in such a way at any other time. It is said that we are all capable of murder, or at least killing, given the right circumstances, and I have no doubt that this is perfectly true. A very good man I know once told me that he could easily have assaulted and killed a man who had knocked down his wife in a motoring accident. My friend had initially thought that his wife had been killed and it was only the realization that she was still alive that stayed his hand. When people are tired or afraid or angry they may well do things they would not do at other times. Of course, this has long been known, not only to people in general, but also in legal contexts in such things as crimes of passion or killing in self-defence.

Personal circumstances and physiological states do not account for the bulk of serious crime. Psychological conditions – mental illness – may well affect people's behaviour adversely toward criminality, but here again we are dealing with individual cases. These potential causes – physical and mental states and illnesses – cannot explain the general problem of crime, nor can they explain why violent crime has increased so markedly in recent years.

It seems clear that the causes of most crimes are moral, social and economic, rather than physical and psychological. There is a strong correlation (dreaded word!) between the recent rise in the crime rate in the West and the great changes in social structure and moral standards that have taken place at the same time. Of course, a correlation does not necessarily imply cause and effect, as we have seen, but, with William of Occam in mind, it is a useful starting point.

POVERTY

Let us look first at poverty, a social condition deemed by many to be an important cause of crime. During the past thirty years or so, a substantial section of the British population became much poorer than it used to be. Much the same happened throughout the English-speaking western world. Such poverty, however it came about, had not been experienced in these countries since the end of the Second World War. At the same time, crime increased sharply. Incidentally, there is no doubt about this; crime has increased in recent years. It is not, as is often claimed, an increase in awareness or greater efficiency in record-keeping that gives the impression that the crime rate has risen, although these things may have played a small part in the matter. There is a great deal of evidence for this rise in the crime rate. For example, the number of murders in England and Wales rose considerably in the period in question. A murder is not the kind of crime that can be overlooked easily; the recorded increase in murders is not a matter that can be explained away by saying that people are now more aware of such things than they used to be, or that police officers simply record more of them. Murder is a horrific event and always has been and awareness of it is nothing new.

To return, after this brief digression, to the role of poverty. Does the correlation between poverty and rising crime in recent years suggest that the former caused the latter? (There is no reason to suppose that the causation is the other way round.) To find an answer, let us look at other places and other times. Many other countries are far, far poorer than Britain and other developed western countries, yet the crime rate in these countries is low, or at least not as high as one would expect. In Britain, the 1920s saw much greater poverty than that experienced today, yet there was hardly any increase in the crime rate then. This kind of evidence seems to suggest that poverty is not an important cause of crime.

I think that such a conclusion is only half true. Dire poverty, such as when a family is in danger of starving to death, would almost certainly cause crime. Any mother would steal to feed her child, if she felt that the child's life depended on it; nor is it possible to fault her for doing so. But this state of affairs hardly ever arises in a country like Britain today. It would seem that the role poverty plays in causing crime in the West is that poor people are very much aware of their poverty. Sources of information, accessible even to the poor in the West, are so abundant that it is almost impossible to remain ignorant of what others have and of the enviable way in which they live. The difference in quality of life between rich and poor is very clearly seen, and is seen as an injustice. This engenders resentment and bitterness. Crime is a natural consequence, because it is impelled by two motives: need and retribution.

But what is "need"? What do we really need in order to live reasonably contented lives? Food, clothes, shelter and what else? Books, money to spend on frivolous things, holidays, toys for the children, televisions, computers – are these things necessities? Of course, the answer will differ from person to person. In that sense, the answer does not matter. People need what they think they need; it is a matter of expectation. A man starving in a third world country would not even aspire to the privileged position of a poor man in London or New York.

It would seem, then, that it is not so much poverty that causes crime, but the acute awareness of one's own poverty, coupled with an acute awareness of other people's riches. People living in dire poverty in Africa have neither the time nor the opportunity to pay much attention to what others have; they are too busy surviving. In the West, poor people still have a good deal of leisure time to think about their own position in society as a whole.

If this is correct – that it is the perception of injustice, resulting from poverty, that causes crime – what can be done about it? This is a very difficult question. In principle, it is easy to say that measures should be taken to alleviate poverty and remove the feeling of isolation and injustice that it engenders. In practice, it is very difficult to see how this can be done. Measures that alleviate poverty do not make poor people rich, they simply make them somewhat less poor. But it is doubtful whether this would make poor people who have turned to crime happier; once the perception of injustice has been aroused, it is very difficult to extinguish.

We started this discussion of poverty with the assumption that, because the rise in poverty went hand in hand with the rise in crime, then some poor people must have turned to crime. It is probably true that some did, indeed, react in this way. But it must still be true that most of the poor are law-abiding. What does not seem to be discussed in the context of crime causation is the question of whether the rich have also turned to crime in the same thirty-year period.

In fact, there is evidence to suggest that this is the case. News about fraud in business, bribery and sleaze in public life, and dubious dealings of all kinds among the wealthy and privileged seems to be much commoner now than it used to be. If this is true, and that this is not only a greater readiness on the part of the press to report such matters, then the recent rise in crime cannot be attributed to poverty alone. Some other influences must be at work.

What could these other influences be? The ebb of family values, the rise in the divorce rate, the impact of television and the decline in religion are often mentioned as causes, if not directly of crime, at least of the erosion of values that used to inhibit crime. Are these changes responsible?

Take religion first. Events in other countries show that the decline in religion may lead to crime and general social irresponsibility. This happened in the Soviet Union, when religion was banned from the schools. It also happened in other communist countries. But it is not possible to conclude that it was general communist oppression that led to this breakdown, since such things happened when religion was abolished by regimes that were not communist, as when the Turkish government banned religious instruction from schools in the late 1920s. In the 1930s, vandalism, drunkenness and a rise in crime were evident in the country's larger towns and cities.

I am not evangelizing. It does not matter in this context what one's personal religious convictions are; it is a fact that the erosion of religious values tends to lead to disorder and chaos. Religion is useful to the magistrate, as Edward Gibbon said. This fact has to be taken very seriously if the problem of crime is to be addressed effectively.

THE FAMILY

Now we come to the question of marriage, divorce and family life. The change of attitude toward marriage and family life has resulted in the generation of new and different priorities among many people. The breakdown of so many marriages has also gone hand in hand with the rise in crime. Britain has one of the highest divorce rates in Europe (much higher than France, Germany or Italy, for example) and it is a country with a rising crime rate. Is it not reasonable to conclude that the breakdown in family life leads to delinquency, and eventually criminality, among the young?

Yes, it is reasonable, but not in the simplistic way in which this problem has often been presented. When a divorce or family break-up takes place, the children's familiar world loses its integrity, forcing them to change their way of life long before they are ready to do so. The main effect that divorce or separation has on children, however, is not so much connected to the separation itself, but the cause and manner of the separation. We are dealing here with a situation in which the observed facts, the breakdown and the children's unhappiness, are probably not cause and effect, but consequences of something else. The divorce or separation of parents usually has a very bad effect on children. Some interesting studies have shed a great deal of light on this important problem, although these findings do not seem to have been appreciated by most politicians who claim to be interested in family values.

The cause of childhood unhappiness is often the series of events that led up to the divorce, not the divorce itself. Disputes between father and mother, especially if they are very acrimonious and bitter, lead to tension in the home. The divorce that follows is also acrimonious, leaving the children confused and unhappy. However, this does not mean that the children of divorced parents necessarily have to suffer misery and confusion. When the cause of the strife is due largely to the behaviour of one parent, the departure of that parent often leads to an improvement in the children's state of mind, since the cause of tension is removed.

In fact, the absence of a divorce in homes in which there is tension between the parents can often have much more harmful results than those that would follow were the marriage to be terminated. It is now becoming clear that one of the most important causes of delinquency and crime among youngsters is tension in the home. An unstable and highly emotional mother or an unpredictable and violent father can unsettle children. Often, the children will emulate one or other of the parents. A mother who throws tantrums or a father who is violent may induce fear in children, who will nevertheless imitate them, the parents' standards being unquestioningly regarded as the norm by most children. This behaviour is then taken into the outside world, with disastrous results.

So, it is not so much the broken home, but the tense home, that puts children at risk of becoming criminally delinquent. It is hardly worth repeating that people and families are subjected to very great pressures these days. These stresses and strains exist in spite of the greatly increased affluence of most people in the West. It is quite probable that this increased affluence, far from relieving tension, actually generates it. It often seems as though affluence has brought with it disillusion. The higher standard of living does not seem to satisfy, after all.

This is not a moral tract, but a book about evidence relating to crime. My purpose is simply to explore the possible causes of crime and it seems to me that if this problem is to be addressed seriously and effectively, then the undoubted increase in the causes of tension within families and within society in general must be examined.

TELEVISUAL MEDIA

Television and other visual media, such as the cinema and the Internet, must have an effect on the behaviour of children. This is often denied, the statement being made that there is no evidence for it. This is not true. Two sources of evidence exist. First, there can be no doubt that our environment has a strong influence on our behaviour; this can hardly be disputed. A child of whatever racial origin will speak English if it is brought up by English-speaking parents; the same child would have spoken Mongolian had it been brought up in Mongolia. Since the evidence for the importance of our surroundings in our behavioural development is so strong, there is no reason to suppose that television is the only part of our environment that does not affect our behaviour. This influence may be either good or bad, of course, but the idea that television influences our lives cannot be dismissed.

The second line of evidence comes from the fact that so many companies and manufacturers pay enormous sums of money to advertise their products or services on television. The aim of an advertisement is to influence the viewers into doing something they would not otherwise have done. Since people do buy the things that they have seen advertised in this way, the inescapable conclusion is that television does influence our behaviour. Indeed, it is difficult to see why television exists at all, if it were not thought to influence behaviour, for, if it did not, what is its purpose?

The real question is whether television influences behaviour adversely or benevolently. The answer, of course, will depend upon the kind of programmes that are watched, the temperament and environment of the child and so forth. Violent programmes might encourage one child to commit violent acts, whereas they might discourage another child from doing so. In short, television is like life: it provides a number of different experiences and it is up to each person what to make of them. Arguments about whether television is harmful or beneficial are, in my view, almost meaningless, since the question seems to boil down to whether one considers "experience" in the abstract to be harmful or beneficial.

THE SOUL OF SOCIETY

I have to say that I believe that far more subtle factors are at work. The mindless, gratuitous crime that is so frequently committed these days must have very deep causes. Crime not for gain, but enacted out of malice and hatred, is something new. The destruction of monuments, such as gravestones and the stained-glass windows of churches, is committed with spiteful glee. Vicious assaults on the very old and the very young used to be taboo in most civilizations, but are today much more common. Some say that such things have always happened, but, in fact, there is little evidence to support that assertion. What is certain is that, when such things happened in the past, they were a cause of deep shame and humiliation to the perpetrator. There is little evidence of such guilty feelings today. One reads daily of the abduction and murder of a child, the burning alive of a young girl, the rape of an old lady. These are worse than crimes; they are acts that lie outside humanity. I know from my own professional experience that, nowadays, the victims of murder are mostly children, old people and women. It is those who are vulnerable who become victims.

Why do these things happen? I trust that the reader does not expect a glib answer, for I do not have one. All I can do is to say that I believe there to be an aching hole in the soul of society, unsatisfied as it is by affluence and material security. This emptiness, if I may be allowed to express myself in a mystical and most unscientific manner, can often be filled by feelings of a very malignant kind. Nowadays it would seem that people do not rage against injustice, but against fellow motorists who get in their way. Disgruntled youths who can find no property to smash and destroy, smash and destroy their own property instead. The urge to inflict damage seems to be irresistible.

My professional involvement in criminal investigation over a period of more than twenty-five years has made me realize that murders, in Britain at least, have little to do with genuine need, or even a burning sense of injustice. I have investigated more than five hundred murders, most of which were the consequences of pure malice and a desire to inflict gratuitous injury and death. Others were the result of sheer envy or low-minded self-indulgence. A garage proprietor kicked to death by two youths because he was Indian; a baby whose skull is smashed to pieces against a wall by the mother's boyfriend; a teenage girl stabbed more than fifty times with a knife by an assailant who felt that she had too many good things; a young girl killed after being seduced by her mother's lover; a young boy murdered by four grown men who felt that killing him after raping him added to their enjoyment ... these are the kinds of crime that I have had to assist in investigating. I have rarely come across a straightforward murder for financial gain or for revenge against a real injustice (not that I condone such crimes, I hasten to add).

Crime is not only a great evil. The way in which it manifests itself – the kinds of crime that are committed – should tell us something about the society in which those crimes have taken place. In the Twentieth Century, Britain was blessed with a lower crime rate than most other European countries, but its crime rate is rising faster than it is in many other countries. The nature of the crimes is also disturbing and confusing.

Let us look at the causes of crime, not only as those things that impel certain individuals to commit them, but as things generated by a civilization as a whole, for history tells us that certain types of crime prevail in certain types of civilization and at certain periods during the growth and decay of that civilization. In any civilization, crimes increase when the purpose of that civilization is forgotten, when the organization of that civilization assumes a greater importance than the civilization itself.

Throughout history, wise and thoughtful people have observed that there are first and second things in life. The second things can be very pleasant; indeed, they may be so pleasant as to appear to be the first things. I think we live in a society that has concentrated too heavily on these second things. If we think more about the first things and less about the second things, if we think more about purposes and less about desires, the causes of crime and other evils may become clearer to us.

CHAPTER ELEVEN

TERRORISM IN THE TWENTY-FIRST CENTURY

If you are a terror to many,
then beware of many
Ausonius

It was 9.15 am on the morning of July 7, 2005. I had a meeting at 9.15 am to attend so and was walking briskly along the corridor to my office after dropping the children at school. As I passed, a colleague rushed out of her room. "Have you heard about the explosions in London?" she cried, "There've been at least three explosions on the Tube near King's Cross during the rush hour!" We headed for my office, anxious to see whether another colleague, who lived in London, had arrived yet. We found her talking on the telephone to her worried husband, reassuring him that she was safe, although unsure if she would be able to get home that evening. Luckily, she had caught an earlier train than normal to get to the meeting on time. During the morning, more information filtered in about the number of casualties and the extent of the damage, as well as news of another explosion, on a bus in Tavistock Square. It soon became clear that, in the wake of the terrible events in the United States America of September 11 2001, our worst fears had been realized: London had once more become a target for terrorist bombs.

Terrorist threats and actions are not new; violent acts for political gain are as old as the formation of political groups themselves. But still, in the past,, terrorists have tended to strike only in their own country or that of its ruler. The post-colonial era, in which the majority of countries have self-rule and the great empires of the past have faded away, has not brought an end to terrorism. On the contrary, new groups and motivations have emerged, and with the dawning of the twenty-first century came the events of 9/11, ushering ushered in a new phase of global terrorism.

Criminals and terrorists now have easier access – through books, television and the Internet – to a wide range of specialist knowledge and sophisticated technology. Their use of these new resources to plan and carry out their crimes/attacks and cover their tracks makes the role of forensic science – the application of any science to matters of the law – more important than ever before.

But there are rays of hope in the fight against terrorism. For instance, on November 8, 2005, the Australian government announced the arrest of 16 suspected potential terrorists, and the seizure of chemicals, weapons and computers after raids on 23 houses in Sydney and Melbourne. This was the culmination of 16 months of investigationin, constituting   one of Australia's largest counter-terrorism operations.

So how exactly can forensic science help in the fight against terrorism? Much of forensic science is about detecting sources: Whose head did this hair come from? Whose blood is this? What does this residue come from? What websites have been visited? It is not, therefore, surprising that forensic science has a vital roles to play in linking evidence left at the scene of a terrorist attack to specific individuals or organizations, as well as in counter-terrorism activities.

WHAT IS TERRORISM?

Before examining the role of forensic science in investigating terrorist attacks, we need to be clear about exactly what we are dealing with. So, what *is* terrorism? The word is derived from the French *terrorisme,* which, in turn, came from the Latin *terrere* (to frighten). *Terrorisme* came to refer to the "Reign of Terror" in Revolutionary France in 1793–94 and the "Reign of Terror" entered the English language in 1795. The rulingFrench government's acts of terrorism consisted of arrests or executions in order to coerce the general public into compliance with its rule. Totalitarian regimes such as Nazi Germany and Soviet Russia similarly employed "terror" and "terrorism" to keep their populations acquiescent. The act of murder for political ends, however, was raised to a fine art by a small group of Ismaili Shi'a Muslims in the eleventh century. These followers of a man known as Hasan-i Sabbah eventually became known as the Assassins. They were renowned for their dagger attacks and the need to protect against such attacks is credited as the driving force behind the development of chain mail.

Today, in most people's minds, terrorism is more broadly defined and is largely associated with political violence, targeted either indiscriminately or specifically aimed at civilians in the absence of a state of war. The terrorists generally do not belong to any recognized armed force. Those accused of being terrorists typically refer to themselves as separatistss, freedom fighters, militants, guerrillas, fedayeen or mujaheddin, terms that refer to their ethnic or ideological struggle. There have been many attempts to develop a more precise definition of terrorism. So, in 2003, the US Department of Defense defined terrorism as: "The calculated use of unlawful violence or threat of unlawful violence to inculcate fear and intended to coerce or to intimidate governments or societies in the pursuit of goals that are generally political, religious, or ideological." The United Nations Office on Drugs and Crime proposed that "[an act of terrorism is] the peacetime equivalent of a war crime". The European Union, in its 2004 definition of terrorism, included "the aim of destabilizing or destroying the fundamental political, constitutional, economic or social structures of a country or international organization". So we can see that the definition of terrorism has evolved in response to political and military circumstances, although there is still no internationally, universally agreed, criminal law definition.

Since, by most definitions, terrorism is used in an attempt to coerce or intimidate a government or society, there are a number of elements that are almost universal in modern terrorist activities. Terrorist attacks are normally regarded – at least by the victim – as "unprovoked" and may well be launched without warning in order to circumvent counter-terrorism measures. Targets often include women and children or are otherwise chosen for maximum propaganda value and to guarantee greater media coverage.

But the scale of terrorist acts varies wildly. The destruction of the World Trade Center in New York on September 11, 2001 must rank as the largest act of terrorism ever, but there have been many other terrorist attacks resulting in significant loss of life and property. We have only to think of events such as the bomb blasts in Bali on October 12, 2002 (when 202 people died), and on October 1, 2005 (when 26 people were killed), the bombing of commuter trains in Madrid on March 11, 2004 (with the loss of 191 lives), the London bombings of July 7, 2005 (with 56 fatalities), and the hotel bomb blasts in Amman, Jordan, on November 9, 2005 (where at least 57 people died), the bomb explosion near shrines in Karbala, Iraq on January 5, 2006 (at least 40 dead), the burning of a local church in Eldoret, Kenya on January 1, 2008 (all 50 members of the congregation perished), and the suicide bomber in Khar city, Pakistan on December 25, 2010 (47 dead and 72 injured) to see that the targeting by terrorists of civilians remains widespread. Yet attacks on military personnel in places where there was no longer any active conflict are also common, such as in Iraq, where, since the end of major military operations in May 2003, suicide bombings, landmines and sniper attacks were part of the daily routine of the foreign military forces that, until December 2011, still occupied that country.

Bombing is now the most common method of modern terrorist attack. Bombs are relatively easy to design, assemble and deliver to the target, and so are particularly attractive to terrorists. Bombing incidents are also sudden and violent in nature, causing high casualty rates and combining the terrorists' twin goals of maximizing publicity and disrupting society. But other publicity-seeking tactics such as kidnapping and hostage-taking are widely employed by modern terrorists. On September 1, 2004, for example, a group of 33 armed men and women – apparently acting on the orders of the Chechen "rebel warlord" Shamil Basayev – stormed Beslan's School Number One in Russia's Ossetia province. They took about 1,300 children, teachers and parents hostage and trip-wires and home-made explosive devices were placed around the gymnasium – where the hostages were held – and the rest of the school. Members of the Russian police, army and special forces surrounded the school, but negotiations failed. Violence erupted on the third day and the school was stormed, with the loss of at least 330 civilians, 186 of them children, as well as all but one of the hostage-takers and 11 Russian soldiers. More recently, in Baghdad, Iraq on October 21, 2010, a group of unknown assailants took the entire congregation of a church hostage, threatening to kill all 120 people if al-Qaeda prisoners were not released. When police attempted to storm the building, the hostage-takers detonated improvised explosive devices in their vests, killing themselves, 46 hostages, and seven policemen.

FORENSIC SCIENCE AND TERRORIST INVESTIGATIONS

But what role does forensic science play in the investigations into terrorist attacks? When a terrorist act has taken place, such as a bomb exploding in a crowded public building, the immediate focus is on the victims. The forensic work may not begin until much later. If there is a fire, it has to be put out. If a bomb was involved, checks must be made to ensure there are no further unexploded devices. Other hazards may include chemicals, radiation hazards, fractured utility mains, damaged power lines and structural damage to the buildings involved. First aid must be given to the wounded and their transport to hospital organized. The safety of any surviving buildings has to be checked. However, almost immediately an investigation will begin, usually posing the questions "who was responsible for this act and how can we identify them?" and "'what methods were used?'", closely followed by "who were the victims?". This is where forensic science has a crucial part to play. To see how forensics can help to answer each of our questions we will use as our examples the Madrid commuter train bombings of 2004, the London Underground bombings in July 2005 and the 9/11 attack on the World Trade Center.

TERRORISTS STRIKE MADRID

It is the little details at a crime scene that can often be very revealing and a bombsite is no exception. In Madrid, between 7.39 am and 7.42 am on March 11, 2004 (the middle of the Spanish rush hour), a co-ordinated series of ten explosions took place on board four commuter trains. It is believed that the explosives had been packed into 13 backpacks or duffle bags, which were then dumped on the four targeted trains. Mobile phones were used as timing devices, but in at least one case the bomb failed to detonate because the mobile phone alarm that should have triggered it was set 12 hours late. A total of 191 people eventually died as a result of the attack, 177 at the scene, and a further 1,800 were wounded. As for any crime, it was vitally important to secure the scene and by 8.00 am, while emergency relief workers were arriving, a "cage operation" was begun, designed to prevent terrorists fleeing the city. By 8.45 am all rail traffic in or out of Madrid was halted, the main Madrid railway stations were closed and parts of the Madrid metro were shut down. Police sealed off streets around the bombsites and evacuated Atocha, the main station that had been targeted. Forensic scientists could then begin their painstaking search for evidence.

But *who* was responsible? The initial prime suspects were the Basque separatist group ETA (Euskadi Ta Askatasuna: Basque Fatherland and Liberation). This group, however, denied responsibility and, indeed, the attack did not bear their usual hallmarks. There had been no warning issued and the scale of the bombing was far greater than anything ETA had previously attempted. Careful forensic investigation of the various bombsites and the three improvised explosive devices that had failed to detonate turned up some interesting pieces of evidence. Forensic analysis of the explosive used showed it to be Goma-2 Eco, an explosive manufactured in Spain and commonly used in the mining industry, but not employed by ETA for some years. A van was found parked outside the station of Alcalá de Henares at the beginning of the commuter line that was bombed. It contained mobile phones and a bag – from which two fingerprints were lifted – containing seven detonators, as well as commercially available audio tapes with Qur'anic verses, a clue that was taken to indicate that an Islamic group rather than ETA might be responsible. Police also investigated reports that at least three men in ski masks had got on and off the various trains at Alcalá de Henares. The investigation into the bombings led within a fortnight to the arrests of at least 14 people, most of them Moroccans, allegedly belonging to the Moroccan Islamic Combatant Group. International arrest warrants were issued for further suspects, several of whom had possible links to the Islamic group al-Qaeda. One of these was linked to the Madrid scene by a mobile phone attached to one of the unexploded bombs, while the two fingerprints found on the detonator bag were eventually identified as being the prints of the middle finger and thumb of Daoud Ouhnane, an Algerian, who is featured on Interpol's Wanted List on behalf of the by Spanish police for questioning in relation to the Madrid attacks.

The Madrid bombing took place three days before Spain's general elections and may have been intended to influence its outcome and so change Spanish policy on Iraq. The Socialist Party had been trailing the ruling Conservatives by a narrow margin in opinion polls, but reaction to the bombing produced a shock change of government, with the Socialists winning 43 per cent of the vote to the Conservatives' 38 per cent. The election is seen as a clear example of terrorism directly affecting a national election and causing a change in policy: soon after his victory, the new Spanish Prime Minister, José Luis Rodríguez Zapatero, reversed the outgoing government's support for the US coalition force in Iraq and announced that Spanish troops would be pulled out from there "as soon as possible".

The message seemed to be that governments backing the United States in the war on Iraq and particularly those such as Spain, the United Kingdom, Australia and Poland who had sent troops, faced the risk of attacks intended to swing public opinion against their governments.

THE LONDON BOMBINGS

From a forensic point of view, the main difference between a "normal" crime scene and one resulting from a terrorist attack is usually one of scale. Let us now examine how forensic science can help to answer the question of *how* a terrorist act was committed.

At 8.50 am on July 7, 2005, during the morning rush-hour, three bombs exploded within 50 seconds of each other on three London Underground trains: the first was on train 204 travelling east on the Circle line between Liverpool Street and Aldgate; the second was on board train 216 travelling west on the Circle line between Edgware Road and Paddington; while the third exploded aboard train 311 on the deep-level southbound Piccadilly line between King's Cross station and Russell Square. Nearly an hour later, at 9.47 am, a fourth bomb exploded aboard a number 30 bus in Tavistock Square. In all 56 people were killed (including the four suspected bombers), and around 700 were injured.

Much later, on September 1, 2005, the terrorist organization al-Qaeda claimed responsibility, but in the meantime investigators needed to find out who the bombers were, whether these were suicide attacks and whether the bombers had acted alone. Once the emergency services had arrived and survivors had been taken to safety, recovery teams and investigators began work at the four sites. The Piccadilly site lay 21.3 metres below ground and posed special problems because it was a much more restricted site. Intense heat of up to 60°C, dust, fumes, asbestos, vermin and fears that the tunnel might collapse, all delayed the extraction of bodies and the forensic operation.

The initial priority, which applies to any explosion, whether from a bomb or a gas leak, is to make sure the scene is safe, and that no further explosions will occur from unexploded devices or any other means. The "forensic" priority is to protect the scene. This latter job is much simpler if the "first responders" i.e. police, fire and ambulance personnel, have had training in the protocols to be observed when attending an explosion scene. If vital evidence is not to be lost, the site must be secured as quickly as possible, a task often complicated by the sheer number and variety of agencies that normally respond. At the Tavistock Square bus-bombing scene, forensic teams were quickly able to contain the seat of the explosion, and police cordoned off a large area. Since debris from a blast can be hurled considerable distances, the rule of thumb is for forensic teams to conduct their search in a circle around the blast up to the point where the furthest fragment from the blast is found, and then widen the radius of the search area by at least 20 per cent, shown visually in this diagram.

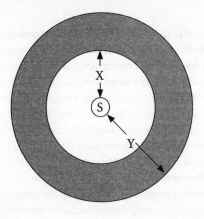

S = seat of explosion
X= furthest dstance that fragments were found from seat
Y = X + 20%X

In Tavistock Square, forensic teams combed, by hand, an area with a radius of approximately 150 metres from the seat and then widened it by 50 per cent, or around another 80 metres. Not only were the streets surrounding the explosion searched but rooftops in the entire area were also combed.

Personnel with previous experience of dealing with a bombsite will have a good knowledge of what they are looking for and so produce the best results. The sorts of forensic evidence commonly found (whole or in pieces) at the site of an explosion include: power sources, varying in size from car batteries to Polaroid film batteries; timers – electronic, mechanical and chemical; switches for arming the explosive device; detonators; printed circuit boards; igniters; explosive charges (often identified from wrapper fragments or residues); wires; adhesive tape; fragments of the containers used to transport or hide the explosive; and other items of bomb-making equipment, such as rubber gloves or rolls of tape, not to mention the explosives themselves. Meticulous records are essential and the search for forensic evidence should be continued until the entire blast area has been thoroughly covered, often a long and labour-intensive process. The search methods used will depend on the number of searchers available, the size and physical layout of the blast area and the degree of destruction caused by the device. Searches are conducted using grid systems, spiral searches, quadrant or zone searches and strip or line searches and every piece of relevant material has to be catalogued and photographed *in situ* before being removed for further examination. It is essential that the scene should be recorded by both sketches and photography and, in large blast sites, aerial photography can also be useful.

Contamination is a particular problem at blast sites, mainly due to the large number of people and variety of agencies that respond. In particular, because military agencies are often involved, care must be taken to ensure that any evidence relating to explosives found at the scene – and needing further analysis – is not handled or transported by any person or vehicle with previous recent exposure to explosives. Forensic scientists must wear protective clothing and caps to avoid leaving any fibres from their own clothes or any hairs at the scene. They wear facemasks to avoid leaving their own DNA at the scene and to protect themselves from breathing in any harmful airborne particles. This latter is particularly important at bomb sites where there will have been mass destruction and upheaval of a wide variety of materials. Latex gloves also serve for general protection and to avoid contamination of evidence.

After the London bombings, forensic analysis of debris from the four sites concluded that each bomb had consisted of about 4.5kg (10 lbs) of home-made peroxide-based explosives packed into a rucksack. No forensic evidence of any timers has been made public, and it has been concluded that the bombs were probably detonated manually.

Israel has suffered from hundreds of suicide bomb attacks and from these the "forensic footprint" of a suicide attack is well documented. In an explosion, gas can expand at up to 20,000 metres per second. All flesh within the immediate explosion zone, including the suicide bomber's body, is pulverized. But sometimes the outer limbs remain intact and, more horrifically, the most obvious sign of a suicide attack is the bomber's head, which is shot into the air at the moment of death. Forensic pathologists carried out a particularly careful examination of two of the bodies from the bus, as evidence suggested that they had been either holding the bomb or sitting very near to it. The finding of a decapitated head at the Tavistock Square site became a major clue that the London bomb attacks might, indeed, be the work of suicide bombers.

More than 800 police officers joined the enquiry and film from 2,500 CCTV cameras in central London alone was examined. Three British-born Pakistani men living in the Leeds area and a British-born Jamaican from Aylesbury who had converted to Islam were eventually named as the bombers. The first three were found to have travelled from Leeds to Luton by car on July 7. There, they were joined by the fourth and they all then boarded a train for London. More explosives, including a nail bomb and detonators were later found in the abandoned car. Raids on six houses in Leeds and Dewsbury led to the seizure of more explosives.

On July 21, 2005, four men tried unsuccessfully to bomb three underground trains and a bus in London. It seems that although the detonators went off, the explosives failed to detonate. Although the men fled the scene, all four were eventually arrested and charged, including one who had fled to Italy. Police recovered four rucksacks and bags left behind by the suspected bombers plus another abandoned in bushes in north London. Forensic experts analysed the failed bombs and found chemical links to the bombs used on July 7.

Following the arrests in Australia in November 2005, mentioned at the start of this chapter, forensic experts found that the explosives seized in those raids were of the same type as those used in the London bombings.

IDENTIFYING THE 9/11 VICTIMS

To see how forensic science can help to answer our third question, "*Who are the victims?*" let us look at some of the events of 9/11.

American Airlines Flight 11 took off from Boston's Logan airport at 7.59 am Eastern Daylight Time (EDT) on Tuesday September 11, 2001. At 8.13 am, the pilots of the Boeing 767 passenger jet made their last contact with Boston Air Traffic Control (ATC), who tried in vain to contact them again at 8.15 am after the plane failed to respond to an order to climb. At 8.20 am, the transponder signal from Flight 11 ceased transmission and the plane veered dramatically northwards away from its planned westward route, before making a 100-degree turn and heading south for New York. Boston ATC concluded the plane had probably been hijacked. Some 20 minutes later it crashed into the 93rd to 98th floors on the north side of the North Tower of the World Trade Center (WTC) in New York. Seismic records have pin-pointed the exact time of the crash as being 8.46.26 EDT. A gaping hole was torn in the side of the tower and the building began to burn at the impact point. The response of the emergency services was almost instantaneous and police, fire-fighters and ambulance crews, together with television news crews, rushed to the scene.

Meanwhile, back at Logan airport, United Airlines Flight 175 took off at 8.14 am, about 16 minutes later than scheduled. At 8.42 am, it veered off its planned course and its transponder soon ceased transmitting. At approximately 9.03 am EDT, roughly 15 minutes after Flight 11 had crashed into the North Tower, Flight 175 was captured on live television approaching the South Tower of the WTC from the southwest and crashing into the 78th to 84th floors. The building had been partly evacuated already after the first crash, so fewer people were trapped above the crashlinecrash line than in the North Tower. The first priorities were to continue to the evacuation of the buildings as quickly as possible and to put out the fires raging around the buildings. However, at about 10.00 am, the South Tower collapsed onto the street below, sending out a massive cloud of dust and debris, and killing all those who had not yet managed to escape the building, including a large number of fire-fighters and police officers. Miraculously, 18 people who had been trapped above the crash line managed to escape. Soon afterwards, at 10.28 am, the North Tower collapsed, again releasing a huge cloud of dust and debris and killing any remaining survivors as well as a large number of rescue personnel. Later that day another one of the seven buildings forming the WTC complex also collapsed. In the case of the WTC attack, the terrorists responsible were identified by the FBI from intelligence information gained from interviews, witnesses, flight manifest logs and passports found at the crash sites. The terrorists' method was simple; they turned two passenger jets into gigantic flying bombs. Forensic science was not really needed to answer the questions of who did it and how, but it was crucial in answering who the victims were.

At the WTC, hundreds of fire-fighters, many off duty, had rushed to the scene to help to contain the fires and evacuate the walking wounded. The Emergency Medical Services, in the form of ambulance crews and triage teams, were summoned and there were also police officers, FBI agents and news crews present. Added to this, were thousands of people, many quite badly injured, trying to flee the scene, making "'securing the scene'" nearly impossible. Further complications for the collection of forensic evidence were the fires that were raging and the sheer scale of the destruction and the resulting spread of debris.

So who were the victims? In the WTC attack of 9/11, all 157 passengers and crew on board the two aircraft were killed, along with an estimated 2,600 people on the ground. It is believed that there were no survivors among the estimated 1,366 people in the North Tower above Floor 93 of the 110-storey building. Television reporters captured horrific scenes of at least 200 people leaping from the Towers to escape the fires raging within. Within two hours of the first attack, both North and South Towers of the

WTC had collapsed into a huge pile of debris, ultimately becoming known as Ground Zero. Many of the dead were buried under tons of pulverized steel and concrete. Hundreds of police and volunteers worked round the clock for weeks to comb through thousands of tons of debris in the search for their remains. By mid-October 2001, workers had collected over 5,000 samples, about a quarter of the final total. The intensity of the fire, together with the collapse of the buildings and the volume of water poured onto the site, meant that many of the remains recovered were very badly degraded, making the task of identification even harder. Almost immediately after 9/11, debris from Ground Zero was shipped to a Staten Island landfill, the Fresh Kills site, and spread out. Trained search and rescue dogs and anthropologists were then brought in to search for further human remains – mostly bone – personal effects and criminal evidence. In all, some 1.62 million tons of material were examined. The last truckload of debris arrived at the landfill site on June 28, 2002 and the search officially ended on July 15, 2002 when the Army Corps of Engineers began dismantling the conveyer belts and screening equipment. At the height of the recovery operation, more than 1,000 workers processed 7,000 tons of rubble a day. The conveyor belts were manned by forensic scientists in respirators, ready to stop the flow of debris if they spotted a bone shard. Other workers used fireworks to keep seagulls and vultures from scavenging the site. After nearly a year of intensive searching, only 292 full bodies or torsos had been recovered from Ground Zero. A total of around 19,916 other specimens were sent to the New York Medical Examiner's office, 6,000 of them small enough to fit into a test tube. Nevertheless, 10,190 body parts, some as small as a fingertip, were eventually identified, primarily through DNA testing. As well as the human remains, boxfuls of rings, watches, wallets and identity cards were accumulated at the Staten Island site.

The task of identifying the victims was helped by the fact that passenger lists existed for the planes and companies within the WTC had payroll records. The New York Police and Fire Departments were also able to provide lists of their missing. For many victims, records were completely lost or never existed, but friends and relatives reported them as missing. Having an identity to work with makes the task of the forensic scientist easier. Traditional techniques involving dental records, fingerprints and identifying features were used for some of the 9/11 victims. Current research in this area includes work at West Virginia University where researchers have been working with the FBI to develop software that will automatically search dental X-rays, speeding the process greatly in dealing with mass disasters.

However, throughout, DNA has played an enormous role, not just in identifying the victims but also in piecing together the various remains to get an accurate full body count. In the Madrid train bombings, for example, the initial death toll of 202 was later revised down because of mis-identification of body parts. This work of identification is slow and painstaking, especially when there is little information about the victim's identity. For the dead of 9/11 a massive DNA identification effort was launched. The New York City authorities held a co-ordination summit on October 3, 2001 to facilitate collaboration between city medical examiners, private gene-sequencing companies, government labs and FBI software specialists. DNA was extracted from remains and sent to companies for matching against victims' relatives, as well as samples from known belongings of some of the dead. The information from these various sources was to be combined into a massive set of databases, which would then use software based on the FBI's felon and population statistics programs in the hope of achieving a match. In the lab, two main techniques were used to match DNA. If the sample contained a cell with an intact nucleus, extraction of the DNA was relatively straightforward. Obtaining a match is much more difficult if the nucleus is degraded, as a result of heat, pressure or dampness – all conditions that affected the samples from 9/11. In these cases, researchers had to extract the DNA from the more durable mitochondria, a more time-consuming process. In many of the bone samples recovered this was the only DNA available. Initially, optimism was high and it was hoped that the majority of victims would be identified. The reality was different and on February 23, 2005, New York officials announced that the process of identifying the human remains from 9/11 would come to an end.

At that point, only 58 per cent of the victims (1,585 people) had been identified on the basis of recovered physical remains and about 9,726 recovered remains were still unidentified. Of the suspected ten hijackers, only two sets of remains had been identified. Since 2006, as work has proceeded to demolish some of the buildings surrounding the WTC site and to build a lasting Memorial, more fragments of human remains have been recovered and, by May 2011, a further 45 victims had been identified, while of the 8,157 bone fragments that require DNA testing, 1,722 have been linked to known individuals, some of whom have already been previously identified as a victim. On September 11, 2011 the National

September 11 Memorial & Museum, on the WTC site, was officially dedicated in a ceremony for family members of the nearly 3000 people killed at the WTC site, at the Pentagon and on board Flight 93 on September 11, 2001, as well as the six people killed in a bombing in the WTC in February 1993. The Memorial consists of two, one acre, reflecting pools sitting in the footprints of the North and South WTC towers, each edged with bronze panels on which are inscribed the names of all who died. The construction of the remainder of the Memorial Plaza and Museum is ongoing, and the Office of the Chief Medical Examiner of New York (OCME) has pledged to continue to attempt DNA identifications of unidentified remains. When the Memorial is complete, it will include a repository at the bedrock level for any unidentified remains. Currently, approximately 40 per cent of all remains found still require testing. This can often be the case even when researchers have much more to go on than a fragment of DNA.

On November 18, 1987 a fire broke out in King's Cross Underground Station in London, claiming the lives of 31 people. Yet it took until January 2004, more than 16 years after the event, before one badly burned male victim, known until then only as victim 115 (the number on the body tag attached to him in the mortuary) was finally officially identified as 72-year-old Alexander Fallon from Falkirk. Mr Fallon had been living rough in London at the time of the fire, although he had kept in regular touch with his four daughters through phone calls and letters. These ceased after the fire and his benefits went unclaimed. Police believed the man they were trying to identify was aged between 40 and 60 years and they had followed up hundreds of enquiries from all over the world relating to missing people thought to have been in London at the time of the fire. A model of the man's face was built up based on the skull from his remains using facial reconstruction techniques. In 2002, Mr Fallon's daughters renewed their enquiries about a possible link between their father and victim 115. Forensic experts compared measurements from the skull with known details about Mr Fallon and declared the mystery solved.

Sadly, however, the same cannot be said for more than four out of every ten families affected by the 9/11 attacks. They still have no recovered remains for burial. Most of the missing dead are presumed to lie in the Fresh Kills site on Staten Island. To the anger and dismay of many of these families, New York Mayor Michael Bloomberg announced on August 18, 2005 that no more work would be carried out to recover human remains from the 9/11 debris at the Fresh Kills site.

FORENSICS AND THE PREVENTION OF TERRORISM

A terrorist act normally follows months or even years of preparation by the terrorists. Intelligence services seek to identify these preparations to prevent attacks and to locate potential terrorists. Forensic science skills can contribute significantly to this work. On February 5, 2001, for example, *USA Today*, published an article entitled *"Terrorist instructions hidden online"*, highlighting the *threat* of illicit use of the Internet and e-mail by terrorists and the need to take steps to counter it. In fact a captured al-Qaeda training manual acknowledged the technical superiority of the US security services and advocated low technology forms of "secret" or covert communication such as invisible ink and simple cyphers. Nevertheless, because of the ever more widespread use of computers, forensic computer scientists can play an important role in linking information on a suspect's computer to a terrorist act. We shall look at computer forensics in more detail in the next chapter, but for now, suffice it to say that after the 1993 bombing of the WTC, there were two computers with 20 megabytes of data each to be examined. After 9/11, forensic computer scientists had to sift through 125 terabytes of data! To put that into context, one terabyte (1012 bytes) is equivalent to 1024 gigabytes. An ordinary personal computer hard drive can hold 400 gigabytes, so 125 terabytes represents more than 300 hard drives filled with information, or around 3,000,000 times as much data as had to be analysed in 1993.

Yet the search can yield results. In August 2004, a year before the London Underground bombings, an alleged al-Qaeda cell based in Luton was broken up after the arrest, in Pakistan, of a known al-Qaeda agent, Mohammad Naeem Noor Khan. It was said that his laptop contained plans for attacks on financial buildings in Washington and New York, as well as attacks on the London Underground.

From a terrorist's point of view, any technique that enables the transfer of information between members of the group in such a way that it is undetectable by anyone outside that group is invaluable for planning and executing a potential attack. For thousands of years codes have been used to provide privacy for a message. Another technique, known as steganography has been used to conceal the very existence of that message. For example, if you are making a purchase over the Internet using your credit card, you do not want your card number to become public knowledge, so a coded message of seeming gibberish is sent which only the seller's web-site can decipher. But although the code may be unbreakable, it is still clear that you have sent a message. For true secrecy, you want to conceal the fact that a message has been sent at all. This is where steganography comes in.

STENANOGRAPHY

The word steganography comes from the Greek words *steganos* (covered) and *graphein* (to write) and literally means "covered writing". It is a technique for concealing the existence of information in such a way that only the intended recipient knows that the hidden message even exists. One of the earliest documents to describe the use of steganography is Herodotus's *Histories*, which dates from around 440 BC. Herodotus tells the story of Demeratus, a Greek living in exile in Persia, who wanted to warn Sparta that the Persian king Xerxes intended to invade Greece. Demeratus scraped the wax off a wooden writing-tablet, engraved the message directly onto the wood and then covered it again with wax. He then sent the apparently blank tablet to Sparta without arousing the suspicions of any of the soldiers along the way.

Modern steganography draws on a wide array of secret communication methods ranging from invisible inks, microdots and character arrangement (distinct from the methods used in cryptography of permutation and substitution), to more sophisticated digital watermarking and computerized image steganography.

As an example, the following message was actually sent by a German spy during World War II: "Apparently neutral's protest is thoroughly discounted and ignored. Isman hard hit. Blockade issue affects pretext for embargo on by-products, ejecting suets and vegetable oils."

If you take the second letter from each word – *pershingsailsfromnyjuni* – then the following message emerges: "Pershing sails from NY June 1."

Computerized applications of steganography have breathtaking implications, including the hiding of large data files inside digital graphic or audio files, or even on ordinary-looking and sounding CD-ROMs, DVDs and digital audio tapes. Commercial applications for the film, entertainment and publishing industries have focused on the prevention of piracy or illegal copying by hiding a "digital watermark" on the CD or DVD in question.

Steganography could clearly play a useful role in the world of espionage since anyone monitoring your communications can probably tell if a message has been encrypted, even if they can't crack the code, but using the best steganographic techniques should mean that a "watcher" does not even know that a message has been sent.

There are now several commercially available steganographic software packages, such as Hide and Seek, StegoDos, S-Tools for Windows and White Noise Storm. These packages exploit the fact that, in digital files in particular, there exists white noise or random information. It is apparently perfectly possible to hide the entire computerized plans for a Stealth Bomber on an ordinary digital audiotape, in addition to its advertised music, by replacing the white noise with a secret message. Anyone playing the tape would simply hear the music.

This disappearing act works by taking advantage of the imprecision of the human ear. Digital music is recorded in 16-bit blocks, but the sound information stored on the sixteenth bit is beyond the normal perception of human hearing. So, if that bit is deleted and replaced with the Stealth Bomber plans, it would be undetectable to the ear. The same logic applies to steganographic algorithms that store data in picture files. Digitized photographs are stored as pixels, which are essentially an array of coloured dots. Each pixel has three numbers associated with it, one for the intensity of each of the colours red, green and blue. Each of these three numbers is stored as eight bits, using a binary code of 0 and 1. The most significant bit (on the left) is worth 128, the remaining bits being worth 64, 32, 16, 8, 4, 2 and finally 1, the least significant bit, on the right. The intensities can range from 0 to 255. Thus a particular shade of green might be represented as:

		128	64	32	16	8	4	2	1
red intensity of 87	=	0	1	0	1	0	1	1	1
green intensity of 128	=	1	0	0	0	0	0	0	0
blue intensity of 52	=	0	0	1	1	0	1	0	0

Hiding other information in the four least significant bits, so that the colour intensities are varied by up to 16 points, has been shown to not affect the original picture's appearance to the human eye. Since text is usually stored with 8 bits per letter, each pixel could hide 1.5 letters. So a standard 640 by 480 pixel image could conceal over 400,000 characters, more than is needed for a standard length novel and plenty of space to hide another image.

After 2001 there were rumours that members of al-Qaeda were doing just this and sending encrypted messages by means of files hidden within digital photographs that were being posted on the auction site eBay. There has never been any official confirmation as to whether this in fact took place.

Fortunately, for every clever method and tool being developed to hide information in multimedia data such as digitized photographs, CD-ROMs or digital audiotapes, equally clever methods are being developed by computer forensic scientists to detect the secret information. The European Union-funded CTOSE (Cyber Tools On-line Search for Evidence) project, for example, produced a series of tools to ensure that anyone involved in the collection, analysis, storage and presentation of electronic evidence, would follow consistent and standardised procedures when collecting and preserving evidence, thus making it easier for evidence to be admitted to any court.

BIOTERRORISM

So far, we have looked at what might be termed "conventional" terrorist activities, but in 2001 the threat of bioterrorism reared its head with a campaign in the USA involving the sending of letters laced with anthrax (*Bacillus anthracis*). A number of bacteria, viruses and fungi pose a serious health risk to humans and could threaten food supplies and affect the environment. All, with the exception of smallpox (*Variola major*) are found in nature. Although microbes or their toxins have been used as weapons for centuries, the "anthrax" letters succeeded in generating fear among the general public.

Micro-organisms make particularly good weapons because they can be cheaply grown from a single organism and it can be difficult to tell where that original organism originated. As a result, a new forensic discipline, microbial forensics, is developing. Infectious disease agents from a specific origin have a unique molecular fingerprint that is nigh on impossible to erase. In theory, the genes of these pathogenic agents could be sequenced. This would need to be done for every strain or isolate known and this information, together with data on the origin of that strain or isolate and details of the laboratories which hold legitimate stores of these pathogens, could be stored in large databases and shared. Another avenue of current research aims to measure the mutation rates in micro-organisms. These rates are thought to be slow in the *anthrax bacillus* but so rapid in some viruses, such as influenza, that different sequence variants can be found in the same person! In the USA, the FBI initiated the Scientific Working Group on Microbial Genetics and Forensics (SWGMGF) in July 2002, to bring together people and agencies with expertise in this field, both nationally and internationally, to establish research frameworks and develop guidelines and protocols for the operation of a national microbial forensics system. Some of the many issues discussed included the proper methods of collecting microbial evidence at a scene, how to decide on the best method for analysing such evidence, the development of validation protocols for any new analytical test and the setting of quality control standards so that microbial forensic evidence would be acceptable in court. Furthermore, the SWGMGF recommended establishing a strain repository of pathogens as well as a database that would link pathogen identification data with other information, such as legitimate sources for a particular microbe, previous occurrence at a crime scene, and so forth, which could be used to support the investigations of law enforcement agencies. Over the past few years, whilst there have been many advances in microbial forensics, particularly in the analytical tools available, many challenges remain to its use in the fight against bioterrorism.

PREVENTING FUTURE ATTACKS

Truly international co-operation is the way forward in the fight not only against terrorism but also global crime. But, research and the sharing of results is vital in many areas if forensic scientists are to stay a step ahead.

In 1995, the European Network of Forensic Science Institutes (ENFSI) was founded and is now a forum for sharing information between forensic scientists through regular programmes of annual and triennial meetings, one-day seminars on special topics and regular meetings of its sixteen Working Groups. It has also encouraged the development of networks, such as the Forensic International Network for Explosives Investigation (FINEX), which was founded in April 2004 by the Netherlands Forensic Institute and the United Kingdom Forensic Explosives Laboratory (DSTL).

In 2010, a new multilateral partnership of regional forensic science networks was established. The International Forensic Strategic Alliance (IFSA) is a truly global partnership between networks from North and South America, Europe, Asia, Australia and New Zealand. Its aim is to "create opportunities for strategic collaboration across the global forensic science community".

Conferences such as the ENFSI "Terrorism and Forensic Science" conference in the Hague in May 2005 and the "Forensic Disaster Victim and Thing Identification" meeting in Prague on October 12, 2005, provided forums for the exchange of information on the latest cutting edge research. Furthermore, 2011 saw forensic science issues included for the first time in the Annual Security Research Conference, held in Warsaw on September

19-21. In Britain, we have the National DNA database, a continually growing register of DNA taken from crime scenes and criminals. By November 2005, the database contained more than 3 million DNA profiles from individuals plus around 250,000 profiles from crime scene samples.

All over the world, forensic science is gaining a higher profile, thanks partly to television programmes such as *CSI*, which give a rather glamorous view of the work of a forensic scientist, but also through a growing recognition of the part that can be played in solving crimes by the careful collection and storage of evidence and detailed, accurate forensic analysis.

There is also a much greater emphasis on rigorous training of new forensic scientists to ensure that everyone involved in forensic analysis, whether of explosives, blood, glass, DNA or microbes, is aware of the standard protocols to be used to ensure results can be confidently admitted to a court of law. In the USA, West Virginia University is one of many to provide degree programmes in Forensic and Investigative Sciences for undergraduates, as well as an active research programme through its Forensic Sciences Initiative. In Australia the University of Technology in Sydney opened its Centre for Forensic Science on November 21, 2001. In Britain, 74 universities will offer 346 full-time degree level courses in various aspects of forensic science from 2012, compared with 39 universities offering 220 courses in 2006 and a mere three courses at three universities in 1996. Forensic laboratories will need to have this range of expertise readily available if they are to assist the law enforcement agencies in the fight against terrorism. While future research will focus on the development and improvement of tools such as the "lab on a chip", fast-screening DNA systems, chemical profiling and internationally compatible databases.

Twenty-first century terrorists plot globally but act locally and so may not be known to local law enforcement agencies. In 2001, researchers at the University of Maryland began compiling the Global Terrorism Database (GTD), which currently holds information on approximately 98,000 terrorist events dating from 1970 to 2010 from all round the world, including location, weapons used, target, casualties and, where possible, who claimed responsibility. This data is freely available as an aid to studying and, hopefully, defeating terrorist violence. After the Madrid bombings, Europol reactivated its counter-terrorist task force and created a centre in Brussels for the threat analysis of terrorist activities. Such initiatives mainly involve information from intelligence and security services, but benefit from input from various forensic disciplines. Some countries have long experience with terrorism and their forensic laboratories and law enforcement agencies have developed a specific expertise in dealing with this type of work. ENFSI and other similar organizations can help other institutes without this background to learn from that experience and expertise. Agencies need to work together to establish how forensic institutes can best support police forces or specialized units effectively in combating terrorism. Forensic institutes and agencies also need to develop systems for working with the law enforcement agencies or specialized anti-terrorist units to allow the effective collection and analysis of forensic samples.

The ultimate goal is to prevent a terrorist attack from ever occurring, but if this fails, then the aim is to bring the people responsible to justice. Forensic science can help to achieve this through rigorous examination of the scene, careful collection and analysis of the evidence and sound interpretation of the results.

CHAPTER TWELVE

COMPUTER FORENSICS

Three things cannot long be hidden: the sun, the moon and the truth

Hindu Prince Gautama Siddharta, the founder of Buddhism, 563-483 BC

Sitting in my study at home in Cambridge in 1998, I placed a CD in the CD-ROM drive on my computer and clicked to open it. The CD contained a folder with photographs of a pleasant family garden with lawn and flower beds and a family group standing in front of a pretty ornamental tree. Another image showed a closer view of the tree under whose leafy canopy stood a garden table laid ready for a celebratory meal including an opened bottle of wine sitting in the centre of the table. Next, a close up shot of a half full wine glass complete with small spider. The case was simple, a customer had bought a bottle of wine from his local wine merchant in Inverness to have at a small family garden party, but on pouring out the wine he also exposed a small dead spider. The customer complained to his wine merchant that the spider must have got inside at the bottling plant, since the wine had only been opened and placed on the table a bare few minutes before the family sat down to their meal, and this was clearly an indication that the cleanliness of the plant must be questionable. As it happened, since this was a special celebration, the customer's wife had taken a photo of the specially laid table with her new Cyber-shot digital camera, as well as snapshots of the assembled family in the garden. The camera being handy, the appearance of the spider was also recorded for posterity and added to the digital photo album the wife had later created on the family's home computer. Anxious to defend its reputation, the vintners who had bottled the wine, at a plant in England, sent the unfortunate spider, by post, to me for identification, which revealed the spider was most definitely a garden-dwelling, silk spinning variety that particularly favoured life in trees and would be most unlikely to be found indoors. I requested further details of the "scene" and the vintners asked the family for a copy of their photographs and duly posted the CD-ROM disk to me in the comfort of my own home. From this, it was possible to suggest that the most likely explanation was that the spider had spun a thread from the tree and, by chance, arrived at the top of the opened bottle on the table below and somehow fallen in. I duly wrote my report, dialled up my internet connection and e-mailed it to the relieved vintners, who nevertheless checked all their hygiene protocols and sent the customer a case of wine as a goodwill gesture.

During the twentieth century, electronic and computer technology has pushed ahead in leaps and bounds. For example, the charge-coupled device (CCD), which was first invented in 1969 by George Smith and Willard Boyle, is the image sensor that is at the heart of all digital cameras. The first commercially available electronic still camera was produced by Sony in 1981, but it was not until the early 1990s that the first professional colour digital camera system became available from Kodak. It had a 1.3 megapixel sensor and was aimed at photojournalists rather than the mass consumer market. The first digital camera that could work with a home computer via a serial cable, and which was aimed at the mass consumer, was Apple's QuickTake 100 camera released in 1994. Nowadays, digital cameras with 12 megapixel sensors are commonplace. We have cameras that can record still photos, zoom in, automatically adjust for light levels, focusing distance, record videos with sound, all at the touch of a button. What's more we now have mobile phones that incorporate the same technology and can take pictures and videos of astounding clarity and, since the 1990s, computer technology has grown and developed at an astonishing rate so that now these digital devices can be easily connected to a home computer and from there to the Internet and the rest of the world.

The result of all these technological advances is that the use of computers and other items of electronic gadgetry has become commonplace in everyday life around the globe. In 2001, when the first edition of this book was published, connections to the Internet were still quite commonly made via a dial-up telephone service; nowadays they are usually made via much faster connections such as a broadband service, a wireless (Wi-Fi) network, or even a satellite link. Back in the 1980s a single computer filled an entire room; today a computer can fit in the palm of your hand. More and more devices are being developed that are capable of connecting to the Internet (or example mobile phones, netbooks, iPads and iPods), and of storing electronic data, (for instance memory sticks, flash drives, personal digital assistants [PDAs], digital cameras, MP3 players and e-readers). Nowadays, travellers to far-flung lands may find themselves with only the most basic of creature comforts and no or an intermittant electricity supply, yet even in these remote corners it may be possible to find an "Internet café" or to use a satellite phone to connect to the internet to send or receive information! Today, the family sitting down to celebrate could well take better photographs (in terms of resolution) of their event on a mobile phone and instantly upload them to a social networking site such as Facebook, or they might take a video on a digital camcorder and share it with the world by "posting" it to sites like YouTube .

Not surprisingly then, criminals have made use of this new technology, both as an aid to committing "traditional" crimes including burglary, fraud, arson, murder, and so forth but also for new, computer-specific, crimes such as hacking and denial-of-service. As a result, computers and other digital devices are often connected to all sorts of cases, both criminal and civil, either as a tool that was used in the commission of a crime, as a source of evidence that a crime took place or as the "victim" of a crime. In all cases, such devices have become increasingly important as a source of evidence.

Computer forensics can be defined as "the application of forensic science techniques to computer-based material". Essentially, it is a combination of computer investigation and analysis techniques involving the identification, preservation, recovery, documentation, analysis and interpretation of digital evidence in such a way that it can be presented in a court of law. In particular, the use of computers and other electronic data storage devices leave behind footprints and data trails, if you know where and how to look for them. As with all forensic evidence, the goal is to collect that evidence without in any way damaging or changing it, and maintaining, at all times, a clearly documented chain of custody so that, for instance, it is clear exactly what happened on a particular computer and who was responsible for it. In Britain, computer forensic scientists voluntarily follow guidelines drawn up by the Association of Chief Police Officers (ACPO) in the *Good Practice Guide for Computer-Based Electronic Evidence*, with the aim of ensuring both the authenticity and the integrity of any electronic evidence.

Before we take a more detailed look at the techniques used in computer forensics, let us just review in a little more detail, the sorts of crimes and evidence that we might be looking for. Take burglary, for example. Many people regularly use social networking sites, like Facebook and Twitter to tell their families and friends about everyday events in their lives, about their homes, their recent purchases and of course, their holidays. A would-be burglar needs only have a friend request accepted to be privy to this information. On the face of it, this may seem unlikely, but in fact many people, particularly celebrities, do accept friend (Facebook) or follower

(Twitter) requests from total strangers! Finding that a suspected thief is indeed "friends" on Facebook with a burglary victim, particularly if they are unconnected in everyday life, would be good evidence for pursuing the investigation. However, what if the burglary was committed the "old-fashioned" way by simply watching a house and noting when its occupants' were away? Well, if stolen items subsequently appeared for sale online, perhaps through auction sites such as eBay, the seller can be traced electronically and subsequently questioned. Similarly, in investigations of crimes like murder and arson, investigators may want to know whether the murder method has been researched online; can any signs of an extramarital affair be found in emails, instant messages or posts to chat rooms; do financial records indicate a possible motive for murder or for an arson-for-profit fire; and so forth.

In the United Kingdom, it is illegal to view hardcore pornographic websites (those that depict acts of necrophilia, bestiality or violence that looks life-threatening) or to possess items that are classified as child pornography. Other countries have different laws and as a result such sites are often hosted overseas but can, of course, be accessed from around the globe. Sexual predators also often use the Internet to seek out child victims online, by means of social networking sites, chat rooms and so on, where the predator makes use of his perception that he is anonymous perhaps to pose as a child to make contact and gain the trust of a potential victim. Investigators in such cases will need to be able to reconstruct a history of the suspect's internet use as well as recovering any relevant material such as emails and photographs stored electronically on a computer or any number of other devices. Likewise, people engaged in organised crime and terrorism may store evidence on their computers. For instance, in the case of the London bombings of July 2005, other evidence led police to suspect four men of being involved and their computers were found to have documents and manuals on how to construct bombs that had been downloaded from the Internet.

Computers themselves can also be the victims of crime. Computers are now an integral part of most people's daily lives, particularly in the West where nearly every household owns at least one computer or device capable of connecting to the Internet. Many people use these devices to keep in touch with family and friends, to shop online for goods and services, for entertainment by downloading music and films, and to store all kinds of personal and confidential information. Furthermore, many facets of society depend upon the computers that control much of the critical infrastructure of civilisation including banking and financial matters, transportation, industry and of course law enforcement, where, as we have seen, the use of computers to store vast databases of information on fingerprints, faces, ballistics records, DNA profiles, and so on, has had a great impact on the ability of investigators to link a suspect to a crime. However, this very dependence brings with it security issues regarding digitally stored data, and makes this technology a particularly appealing target for both criminals and would-be terrorists. Computers are vulnerable to unauthorised access to protected data, known as hacker attacks. The most common technique used by hackers is password cracking. Passwords are used for many purposes, and many users do not create very complex passwords, or have trouble remembering more than one, so use the same password for everything. Hackers generally use password-cracking programmes to try to gain unauthorised access to individual accounts, but they may also use other methods, such as spyware, keylogger and sniffer programmes, to hack into websites and gain access to databases for example, in order to illegally copy, alter in some way or even destroy data. There are, for instance, several cases on record of disgruntled employees gaining unauthorised access to a former employer's databases of customer information and deleting them! The Internet is a vast global network that links millions of smaller computer networks and individual computers through the same set of communication protocols, thus allowing the exchange of information, news, opinions and so forth. It, too, is vulnerable to electronic sabotage, for instance Denial of Service (DoS) attacks and the spread of malicious software (malware). In a DoS attack, a particular network server is bombarded with more, usually fake, authentication requests than it can cope with causing it to ultimately shut down, while malware is software designed to invade an individual computer with the aim of either causing damage to that computer, or of stealing information from it. Viruses, worms, Trojan horses and botnets are examples of different forms of malware. As computer security becomes ever more sophisticated with encryption protocols, firewalls and anti-malware applications to prevent unauthorised access, so those who wish to gain access, whether for nefarious intent or simply for "bragging rights", devise new ways of trying to get around them.

Thus, the main emphasis in computer forensics is on data recovery. Key questions are: what data are stored on a particular computer, be that in the form of documents, books, messages, photographs, video clips, and so on, where did that data come from and who put it there. To this end, key skills are knowing the best places to look for evidence and knowing when to stop looking! To do a thorough job, the investigator should know the hardware, operating systems, file systems and networking solutions associated with all the equipment under investigation, be they computers, cameras, mobile phones or any other data storage device. The methods used for extracting data vary depending on the device in question and a detailed study of them all is beyond the scope of this chapter. I will instead confine the remainder of the discussion to the device most commonly examined, the personal computer, however it should be borne in mind that sound forensic practices should always be applied to any investigation of electronic data.

So, first we will look take a look at what exactly we mean by a computer. The physical components of a computer, the chassis, keyboard, monitor, mouse, power cable, hard disk drive (HDD), motherboard, central processing unit (CPU), random-access memory (RAM), read-only memory (ROM), speakers, printer, in short any computer component or peripheral piece of equipment connected to a computer that you can see or touch is referred to as "hardware". "Software" is a set of instructions compiled into programmes and applications that will perform a particular task on the hardware. Examples of software include operating systems such as Windows, Mac OS and Linux, web browsing applications like Firefox, Safari and Internet Explorer, word processing programmes such as Microsoft Word and WordPerfect, accounting programmes including Microsoft Money and Quicken, and statistical analysis programmes like SPSS and Statistica, to name but a few. The operating system is the link between the electronic components of the computer and the human user. Data is usually stored on the HDD by means of a file system, and each operating system generally runs a unique data file system, although some can support the method of others. The HDD is made up of blocks. To store data, the HDD has first to be partitioned, which is done by simply defining a contiguous set of blocks as belonging together; they are then treated as an independent disk. A single HDD can thus hold several partitions, which appear as several disks. The structure of the file system is then created, and this is termed formatting. Some examples of file systems are FAT32 and NTFS (Windows file systems), HPFS (some Mackintosh systems) and EXT3 (Linux systems). Each has its own method of storing, retrieving and allocating data, which the computer forensic specialist must be familiar with.

So how should an electronic crime scene be processed? The answer is, in much the same way as a traditional crime scene, by documenting the scene in as much detail as possible using sketches and photographs. However, as well as getting views of the overall scene, it is also vitally important that close-up photographs are taken of all peripheral devices such as mouse, keyboard, monitor, speakers, etc, their connections to the main system unit and their serial numbers, if practicable. If any monitor is displaying information, this should also be photographed. If the computer is part of a network, then a technical network sketch should also be made. Finally, all cords to peripheral devices should be labelled and a corresponding label attached to the computer port it was connected to so that system setup can be accurately reconstructed in the laboratory or in the courtroom.

Once the scene has been adequately recorded, the examiner needs to physically isolate the computer in question to make sure that it cannot be accidentally contaminated. This may involve running a full system shutdown, or simply pulling the plug from the back of the computer or possibly attempting a "live" acquisition of any data. The final decision depends on many factors: is there an encryption programme running which would make the data unreadable without a key or password if power is cut; is data being deleted which can be halted by pulling the plug; and so on. If the data is on a hardware device such as a CD-ROM, an external hard drive, a USB Flash Drive, a memory card, or stick, such data is generally considered to be "non-volatile", in other words the data will remain on the hardware even if the electricity supply to the device is turned off. In contrast, hardware such as a printer, monitor or keyboard is not normally used to store data, simply to send or receive it, and once such devices are turned off, no information is stored. Or is it? In fact some data may be retrievable, for instance from a printer buffer, even if the device has been completely shut down. In any event, best practice is to secure the computer and its associated peripherals if they are to be used as evidence and take them to a computer forensics laboratory.

Once in the laboratory, the most important task is to obtain whatever relevant data is present for analysis. The examiner must use the least intrusive method at all times so that data can be retrieved from the HDD without altering the smallest particle of it. Because rebooting a HDD to its operating system changes many files and could potentially destroy data that could otherwise be used as evidence, best practice dictates that the HDD is removed from its native system and placed into a laboratory forensic computer so that a digital copy, or forensic image, of the hard drive can be made. Once the original HDD has been copied, it is locked in a safe or other secure storage facility to maintain its pristine condition. All investigations are then carried out on the digital copy.

Electronic data analysis is virtually limitless and really only bound by the familiarity of the examiner with the device or system in question. Hence even a discussion of all of the potential sources of electronic evidence on a personal computer is beyond the scope of this chapter, so we will concentrate on two of the most common areas of analysis, namely visible and latent data. Visible data includes all information that the operating system is aware of and that is easily accessible to the user, so this would include files produced by the computer user, such as letters, spreadsheets, balance sheets, photographs. However, there are other areas where visible data can be found, such as in temporary files, swap files, logs and directories, if one knows where to look. Latent data is data that is hidden in some way (and not necessarily intentionally) from the user's view, for instance when a file is deleted, the data contained within it actually remains in the same location on the HDD until that particular spot is overwritten by new data. Again, latent data can be found in many places, if you know where to look. Investigators use a variety of techniques and specialised forensic applications, such as EnCase or Forensic Toolkit for Windows systems and Forensic Autopsy or SMART for Linux systems, to examine the forensic image of the HDD. These applications search hidden folders, unallocated disk space, slack space, swap space and so forth (at what is termed the binary level) for copies of deleted, encrypted, partially overwritten or damaged files. Any evidence found on the digital copy must be carefully documented and subsequently verified using the original HDD if such evidence is to be used in a court of law.

For the final part of this brief overview of computer forensics, we will consider the forensic analysis of Internet data and communications. As mentioned previously, the Internet is essentially a network of computer networks that allows any computer to "speak" to any other computer provided each one is connected to the Internet. Information can travel over the Internet by means of a variety of computer languages known as protocols, for example email relies on a protocol or language known as SMTP or Simple Mail Transfer Protocol, while NNTP or the Network News Transfer Protocol is used to post, distribute and retrieve news items on the worldwide bulletin board system known as USENET. The World Wide Web (or simply the Web) is not quite the same thing as the Internet. The Web is simply one portion of the Internet, albeit the largest portion, and is essentially a network of computer servers that use HTTP, the Hyper Text Transfer Protocol, to allow applications to communicate and share information that is stored in the form of Web pages. These Web pages may contain text, pictures, graphics, sound and video. Web pages are linked to each other by means of hyperlinks (essentially a sort of bookmark to either link different documents, or to link different parts of the same document) and are usually accessed using Web browsers such as Internet Explorer, Firefox, Safari etc.

Since the Internet and World Wide Web are essentially vast sources of information, it should not be surprising that use of the Internet is often associated with crime; enticing a child to a paedophile ring, planning a robbery, embezzling funds from an employer's business, harassing an acquaintance and so on and so forth. Fortunately for the investigator, there is often a lot of evidence left behind from a user's Internet activity, if you know where to look! Some of the most useful sources of information on internet use for the forensic investigator can be gleaned from the Internet cache, from Cookies, Bookmarks and Web browser's Internet history. Briefly, when a user visits a particular web site, their Web browser may store all or part of that web page in a "cache" on the user's HDD, so that if that particular page is revisited, portions of the page can be reconstructed from the saved data and the whole page can be opened more quickly for the user. A Cookie is a small file that some Web sites place on the local HDD to track information about visitors to that site. Cookies may record number of visits to the site, purchasing habits (if relevant), passwords or even personal information in order to recognise the user next time they visit the site. Cookies can only been placed on the HDD if the Web browser being used, Internet Explorer for example, has been configured to allow this. However, many users store the Web address of their favourite sites in their "bookmarks" or "favourites", which work rather like having a preset TV channel, in that the user simply clicks on the desired bookmark and the relevant Web page will open. Furthermore, all Web browsers track the history of Web page visits for the user, mainly as a matter of convenience, rather like the recent calls list on a mobile phone. Many Web browsers can store up to several weeks' worth of Web sites visited, to enable users to easily revisit sites just by going through their browser history. Some browsers also store information on the actual files that have been accessed over a network, or even from external hardware such as CDs and memory sticks, in the Internet history.

Much can be learned about the interests of the user from studying their bookmarked sites and Internet history, and as we have seen many times already, even if a user attempts to hide their Internet tracks, by deleting the cache for example, there are specialised programmes available to help the investigator recover the missing information.

Finally, access to the Internet opens up a whole new world of possibilities in terms of communication between family, friends, business colleagues and total strangers. These include back-and-forth exchanges via email, instant message conversations, group chats in "chatrooms" and posting messages on forums. All such communications are important in forensic investigations. Fortunately, some rules are necessary for computers to communicate on a global network, whether it be brother and sister sending instant messages to each other from adjacent rooms to corporate deals being brokered in several centres spread out across the globe, from posting messages on sales forums to sell stolen good to paedophiles joining chatrooms aimed at teenages to seek out new victims. All computers connect to the Internet by means of an Internet Service Provider (ISP). The fundamental rule is that any computer that is connected to the Internet must have an address, known as an Internet Protocol (IP) address, from the ISP they use to make their connection, and it is this IP address that provides investigators with a means of identifying a particular user since each ISP should keep records of the link between IP address and real person. Although IP addresses are located in different places for different methods of Internet communication, once again it is usually possible to find the IP address if one knows where to look. The same is true for accessing the actual messages. Emails tend to be stored on the HDD for personal users, while work mail is often stored on a central mail or file server. Many people use free Web-based email services, such as Hotmail, which allow them to remotely access their email from any computer, and this will leave a trail in the Internet cache. By their very nature, instant messaging and chatroom conversations are not intended to be preserved for posterity (although most of the software that runs such services does allow for archiving conversations) so generally speaking conversations of this nature are stored in the random-access memory (RAM) space of a computer, which is a volatile memory space. In other words, RAM only holds data whilst it has power, if the computer is unplugged any data in RAM will be lost – but not completely. Again, with the right tools and knowledge, it is often possible to recover at least fragments, however disconnected and incomplete, of such conversations.

In summary, we live now in an age where computers and other electronic devices capable of storing digital data are a part of our everyday lives. As a result, computers, digital cameras, mobile phones and many other electronic devices often have a part to play in the investigation of a crime. As with all forensic investigations, careful examination and meticulous records together with sound technical knowledge of what to look for and how to find it can yield valuable evidence.